LONDON MATHEMATICAL SOCIETY LECTURE NOTE SERIES

Managing Editor: Professor J.W.S. Cassels, Department of Pure Mathematics and Mathematical Statistics, University of Cambridge, 16 Mill Lane, Cambridge CB2 1SB, England

The books in the series listed below are available from booksellers, or, in case of difficulty, from Cambridge University Press.

D1594618

London Mathematical Society Lecture Note Series. 148

Helices and Vector Bundles: Seminaire Rudakov

A.N. Rudakov;
A.I. Bondal, A.L. Gorodentsev, B.V. Karpov,
M.M. Kapranov, S.A. Kuleshov, A.V. Kvichansky,
D.Yu. Nogin, S.K. Zube.
Moscow State University

Translated by
A.D. King, P. Kobak and A. Maciocia

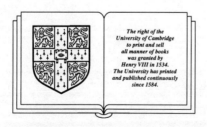

The right of the
University of Cambridge
to print and sell
all manner of books
was granted by
Henry VIII in 1534.
The University has printed
and published continuously
since 1584.

CAMBRIDGE UNIVERSITY PRESS

Cambridge

New York Port Chester Melbourne Sydney

Published by the Press Syndicate of the University of Cambridge
The Pitt Building, Trumpington Street, Cambridge CB2 1RP
40 West 20th Street, New York, NY 10011, USA
10, Stamford Road, Oakleigh, Melbourne 3166, Australia

© Cambridge University Press 1990

First published 1990

Printed in Great Britain at the University Press, Cambridge

Library of Congress cataloguing in publication data available

British Library cataloguing in publication data available

ISBN   0 521 38811 2

Cstack
Scimon

Contents

1. Exceptional Collections, Mutations and Helices

A.N. Rudakov

We shall give a general axiomatic presentation of the theory of helices and introduce some general definitions and notations.

Research on exceptional bundles was started in Moscow University after a lecture given by A.N. Tyurin given in the autumn of 1984 on a preprint of [1]. In that paper a theorem is given describing the possible Chern classes which a stable bundle on P^2 can have. Exceptional bundles appeared as some sort of boundary points. The results of the first one and a half years of our work were presented in [4]. Papers [3] and [6] together with subsequent papers represent the research of the following one and a half years. Most of the papers use a technique which should be called Helix Theory.

The definition of a helix and the first results about helices appeared in [4]. The key lemma 2.2 of that paper and the first version of the definition of a helix are due to Gorodentsev [3]. These constructions were a generalisation to arbitrary dimensions of a method of Rudakov which assigned an exceptional bundle on a projective plane to a pair of exact sequences [5]. The word "helix" and the idea of considering a helix as an infinite system of bundles with some form of periodicity is due to W.N. Danilov.

Further development of the notion of a helix was connected with applications of the primary ideas in new contexts. This was done by Gorodentsev for arbitrary categories of coherent sheaves [3], by Rudakov for a category of symmetric sheaves on a two-dimensional quadric [6] and by others in subsequent papers. The aim of the present paper is to formulate the most general definition of a helix, which will encompass all of the present applications.

1. Helices.

We shall consider pairs of objects of a category $\mathcal{U}$ or elements of a set $\mathcal{U}$.

In both cases we need to distinguish certain pairs. A pair (A, B) is *left admissible* if a certain pair $(L_A B, A)$ is defined and is *right admissible* if the pair $(B, R_B A)$ is defined.

(1L) If (A, B) is left admissible then $(L_A B, A)$ is right admissible and $R_A L_A B = B$.

(1R) If (A, B) is right admissible then $(B, R_B A)$ is left admissible and $L_B R_B A = A$.

If (1L) and (1R) hold then there is a one-one correspondence between left and right admissible pairs, but it does not eliminate the possibility that some pairs may be both right and left admissible.

The pair $(L_A B, A)$ is called the *left mutation* of (A, B), and $(B, R_B A)$ the *right mutation*. We will also call the object $L_A B$ the *left shift* of B, and $R_B A$ the *right shift* of A.

(2L) Let A, B, C be such that the pairs (B, C), $(A, L_B C)$ and (A, B) are left admissible. Then the pairs (A, C), $(B', L_A C)$ are left admissible, where $B' = L_A B$, and $L_A L_B C = L_{B'} L_A C$.

(2R) Let A, B, C be such that the pairs (B, C), $(R_B A, C)$ and (A, B) are left admissible. Then the pairs (A, C), $(R_C A, B')$ are right admissible, where $B' = R_C B$, and $R_C R_B A = R_{B'} R_C A$.

Note that if axioms (1R) and (1L) are satisfied then condition (2L) is equivalent to the following:

(2L) Let A, B, C be such that the pairs (A, B), $(R_B A, C)$ are right admissible and the pair (B, C) is left admissible. Then the pairs (A, C'), $(R_{C'} A, B)$ are right admissible, where $C' = L_B C$, and $R_C R_B A = R_B R_{C'} A$.

It is easy to formulate an analogous condition for (2R). The equalities in Axioms (2L) and (2R) are usually called the *triangle equations*.

It will be convenient to denote the object $L_A L_B C$ which appeared in (2L) by $L^{(2)} C$ and also to set $R^{(2)} A = R_C R_B A$. In the same way, if $A_1, \ldots, A_s$ is a system of objects we put $L^{(0)} A_s = A_s$, $L^{(1)} A_s = L_{A_{s-1}} A_s$, $L^{(i)} A_s = L_{A_{s-i}} L^{(i-1)} A_s$, with the condition that the resulting pairs are left admissible. Analogous notation will be used for right mutations.

DEFINITION. The collection $\{A_i \mid i \in \mathbf{Z}\}$ will be called a *helix of period n* if for all s the following condition is satisfied:

(hel) The pairs (A_{s-1}, A_s), $(A_{s-2}, L^{(1)} A_s)$, $\ldots$, $(A_{s-n+1}, L^{(n-2)} A_s)$ are left admissible and $L^{(n-1)} A_s = A_{s-n}$.

Further on we shall be assuming that (1L), (1R), (2L) and (2R) are satisfied. Then (hel) is equivalent to

(hel)$'$ The pairs (A_{s-n}, A_{s-n+1}), $(R^{(1)} A_{s-n}, A_{s-n+2})$, $\ldots$, $(R^{(n-2)} A_{s-n}, A_s)$ are right admissible, and $R^{(n-1)} A_{s-n} = A_s$.

It often happens that there is an involution $*$ on $\mathcal{U}$; $A^{**} = A$. We will say that the involution $*$ is compatible with mutations if (A, B) is left admissible precisely when (A^*, B^*) is right admissible, and $(L_A B)^* = R_{A^*} B^*$. Given such an involution, theorems concerning left mutations will imply analogous theorems for right mutations and vice-versa.

Each collection of the form A_i, $A_{i+1}, \ldots, A_{i+n}$ is called a *foundation* of the helix $\{A_i\}$. Note that a helix is uniquely determined by any of its foundations.

A collection $\{B_i,\ i \in \mathbf{Z}\}$, with

$$\begin{aligned}
B_i &= LA_{i+1} &&\text{for} \quad i \equiv m - 1 \,(\mathrm{mod}\,n), \\
B_i &= A_{i-1} &&\text{for} \quad i \equiv m \,(\mathrm{mod}\,n), \\
B_i &= A_i &&\text{for} \quad i \not\equiv m, m - 1 \,(\mathrm{mod}\,n),
\end{aligned}$$

is called a *left mutation* of a helix at A_m.

A collection $\{C_i,\ i \in \mathbf{Z}\}$, with

$$\begin{aligned}
C_i &= RA_{i-1} &&\text{for} \quad i \equiv m + 1 \,(\mathrm{mod}\,n), \\
C_i &= A_{i+1} &&\text{for} \quad i \equiv m \,(\mathrm{mod}\,n), \\
C_i &= A_i &&\text{for} \quad i \not\equiv m, m + 1 \,(\mathrm{mod}\,n),
\end{aligned}$$

is called a *right mutation* of a helix at A_m.

The basic fact about helices is the following.

THEOREM. *A right or left mutation of a helix is again a helix.*

All applications of helices are based on this theorem and the fact that any property of some given helix which is preserved by mutations will also hold for any mutation of that helix.

PROOF. (of the theorem) The symmetry of the definitions implies that it is enough to prove the theorem for a left mutation. In the proof we will consider collections $\{A_i \mid i \in \mathbf{Z}\}$ for which (hel) holds for some subset of the indices and not necessarily for them all.

LEMMA 1. *Let $\{A_i\}$ be a collection for which (hel) holds for $s \in S \subset \mathbf{Z}$ and let $\{B_i\}$ be obtained from $\{A_i\}$ by a left mutation at the point A_m, $m \in S$. Then $\{B_i\}$ is a helix for all $s \in \mathcal{L}_m S$, where*

$$\mathcal{L}_m S = \left(S \setminus \{i \mid i \equiv m \,(\mathrm{mod}\,n)\} \right) \cup \left\{ i \mid i \equiv m - 1 \,(\mathrm{mod}\,n) \right\}.$$

PROOF. As the choice of the position of a mutation is determined up to the period n, we can assume that $m - 1 \leqslant s < m + n - 1$. If $s = m - 1$ then

$$B_s = L^{(1)} A_m,$$

$$(B_{s-1}, B_s) = (A_{m-2}, L^{(1)} A_m),$$

$$\vdots$$

$$(B_{s-n+1}, L^{(n-2)} B_s) = (A_{m-n-1}, A_{m-n}).$$

The last equality holds because $L^{(n-2)}B_s = L^{(n-1)}A_m = A_{m-n}$. It is clear that $L^{(n-1)}B_s = L^{(1)}A_{m-n} = B_{s-1}$. We must now prove helicity for all $s \in S$ in the range $m+1 \leqslant s < m+n-1$. We see that $L^{(i)}A_s = L^{(i)}B_s$ for $i \leqslant s-m-1$. Put $A = A_{m-1}$, $B = A_m$, $C = L^{(s-m-1)}A_s$, and apply Axiom (2L). The conditions of the axiom are fulfilled because $s \in S$ and so we have helicity. Then the pairs (A, C) and $(B', L_A C)$ are equal to the pairs $(B_m, L^{(s-m-1)}B_s)$ and $(B_{m-1}, L^{(s-m)}B_s)$ respectively. We can see that they are left admissible and $L^{s-m+1}B_s = L^{s-m+1}A_s$. Now it is easy to see that such an equality is preserved and that all pairs arising from $\{B_i\}$ coincide with those for $\{A_i\}$. As a result $L^{(n-1)}B_s = L^{(n-1)}A_s = A_{s-n} = B_{s-n}$, and the lemma follows.

COROLLARY.

1) *Property (hel) is satisfied for all $s \in S$ if and only if (hel)' is satisfied for all $s \in (S - \{n\})$.*

2) *If the collection $\{C_i\}$ is obtained from $\{A_i\}$ by right mutation at the point A_m, then $\{C_i\}$ satisfies (hel) for all $s \in \mathcal{R}_m S$, where*

$$\mathcal{R}_m S = \left(S \setminus \{i \mid i \equiv m \,(\mathrm{mod}\,n)\}\right) \cup \left\{i \mid i \equiv m+1 \,(\mathrm{mod}\,n)\right\}.$$

LEMMA 2. *If $\{A_i\}$ is a helix, then $\{B_i\}$ satisfies (hel) when $s \equiv m \,(\mathrm{mod}\,n)$.*

PROOF. We can assume that $s = m$. Note that since $m \in S$, the collection $\{B_i\}$ can be obtained from $\{A_i\}$ by the successive application of $n - 2$ right mutations at the points with indices $m - n, m - n + 1, \ldots, m - 2$. Applying part (2) of the corollary we see that m is always left in the set of those indices for which (hel) is satisfied.

To complete the proof of the theorem we note that if $\{A_i\}$ is a helix for $S = \mathbb{Z}$ then, from Lemma 1 the collection $\{B_i\}$ is a helix for $s \not\equiv m \,(\mathrm{mod}\,n)$. But from Lemma 2, (hel) is also satisfied for these s. This proves the theorem.

2. Notation.

In the applications of the theory of helices there are a lot of computations with the functors Hom and Ext^i. In order to write these calculations down in a more compact form, we will use the notation

$$^i\langle A \mid B \rangle = \mathrm{Ext}^i(A, B).$$

Generalizing the traditional notation in physics, we denote the functor $\mathrm{Ext}^i(A, -)$ by $^i\langle A \mid$, and the functor $\mathrm{Ext}^i(-, B)$ by $^i\mid B \rangle$. When applying such notation to the functor Hom we shall drop the "0" whenever this will not lead to confusion.

3. Exceptional Objects.

Let $\mathcal{F}$ be a vector bundle or a sheaf on an n-dimensional variety X over a field k. We call $\mathcal{F}$ *exceptional* if

$$^0\langle \mathcal{F} \mid \mathcal{F} \rangle = k, \quad \text{and} \quad {}^i\langle \mathcal{F} \mid \mathcal{F} \rangle = 0 \quad \text{for} \quad 0 < i < n,$$

and the space $^n\langle \mathcal{F} \mid \mathcal{F} \rangle$ has the smallest possible dimension. More concretely, if X is a projective space, a quadric or any other variety with an ample anticanonical class then we must have $^n\langle \mathcal{F} \mid \mathcal{F} \rangle = 0$. If X is a K3 surface, an elliptic curve or any other variety with zero canonical class, then $^n\langle \mathcal{F} \mid \mathcal{F} \rangle = k$.

Rank one bundles are always exceptional, so we can express the fact that $\mathcal{F}$ is exceptional in the form

$$^i\langle \mathcal{F} \mid \mathcal{F} \rangle = {}^i\langle \mathcal{O}_X \mid \mathcal{O}_X \rangle \qquad \text{for all } i.$$

4. Mutations.

In studying exceptional bundles on varieties with ample anticanonical classes it seems to be convenient to use the notion of an exceptional collection [4].

By an *exceptional collection* we mean a collection of exceptional sheaves $F_1, \ldots, F_s$ on X, such that for all $1 \leqslant l < m \leqslant s$,

$$^i\langle F_m \mid F_l \rangle = 0 \qquad \text{for} \quad i \geqslant 0,$$
$$^i\langle F_l \mid F_m \rangle = 0 \qquad \text{for all } i \geqslant 0, \text{ except for perhaps one value of } i$$

We can now consider mutations of exceptional pairs. Let (A, B) be an exceptional pair. We say that this pair is left admissible with shift $L_A B$ if one of the following three possibilities occur:

(1) $^0\langle A \mid B \rangle \neq 0$ and the canonical map $^0\langle A \mid B \rangle \otimes A \to B$ is epimorphic. Then define $L_A B$ by

$$0 \to L_A B \to {}^0\langle A \mid B \rangle \otimes A \to B \to 0,$$

(2) $^0\langle A \mid B \rangle \neq 0$ and the canonical map $^0\langle A \mid B \rangle \otimes A \to B$ is monomorphic. Then define $L_A B$ by

$$0 \to {}^0\langle A \mid B \rangle \otimes A \to B \to L_A B \to 0.$$

(3) $^1\langle A \mid B \rangle \neq 0$. Then $L_A B$ is defined to be the universal extension

$$0 \to B \to L_A B \to {}^1\langle A \mid B \rangle \otimes A \to 0.$$

Case (1) is called *division*, case (2) is called *recoil* and case (3) is called *extension*.

Similarly, right mutations are defined from the following sequences:

$$0 \to A \to {}^0\langle A \mid B \rangle^* \otimes B \to R_B A \to 0 \qquad \text{division;}$$

$$0 \to R_B A \to A \to {}^0\langle A \mid B \rangle^* \otimes B \to 0 \qquad \text{recoil;}$$

$$0 \to {}^1\langle A \mid B \rangle^* \otimes B \to R_B A \to A \to 0 \qquad \text{extension.}$$

Where the last sequence is the universal extension.

In some cases, for example $X = \mathbf{P}^2$, one can show that all the mutations turn out to be divisions [4]. In the case $X = \mathbf{P}^n$, $n > 2$ in [4], this is made part of the definition and only such mutations are considered. In that paper some theorems are proved which verify the axioms of helix theory. In particular it is shown that the bundles $\{\mathcal{O}_{\mathbf{P}^n}(i)\}$ form a helix.

References

[1] GORODENTSEV, A.L., Transformations of Exceptional Bundles on $\mathbf{P}^n$, *Math. USSR Isv.*, **32** (1989) 1–13.

[2] GORODENTSEV, A.L., Exceptional Bundles on Surfaces with a Moving Anti-canonical Class, *Math. USSR Isv.*, **33** (1989) 67–83.

[3] GORODENTSEV, A.L., & RUDAKOV, A.N., Exceptional Vector Bundles on Projective Space, *Duke Math. J.*, **54** (1987) 115–130.

[4] DREZET, J-M., & LE POTIER, J., Fibrés Stables et Fibrés Exceptionelles sur $\mathbf{P}_2$, *An. Ecole Norm. Sup.*, **18** (1985) 193–244.

[5] RUDAKOV, A.N., The Markov Numbers and Exceptional Bundles on $\mathbf{P}^2$, *Math. USSR Isv.*, **32** (1989) 99–112.

[6] RUDAKOV, A.N., Exceptional Bundles on a Quadric, *Math. USSR Isv.*, **33** (1989) 115–138.

2. Construction of Bundles on an Elliptic Curve

S.A. Kuleshov

M.F. Atiyah [1] proves that the moduli space of indecomposable bundles with a fixed degree and rank over an elliptic curve is isomorphic to the curve itself. Atiyah also shows that an indecomposable bundle over an elliptic curve is simple if and only if the highest common factor of its degree and rank is equal to one.

In this paper Atiyah's results are proved using a recent technique whose power is frequently demonstrated in this collection of papers. We wish to stress that the proof given here of Atiyah's first two theorems (on the moduli space of indecomposable bundles) really only differs from Atiyah's proof in terminology. However, the proof of the theorem on simple bundles is fundamentally different from Atiyah's and the author hopes that it is more elegant.

Apart from proving Atiyah's theorems, this paper will classify the simple bundles whose determinant is equal to a multiple of a fixed point. The classification resembles that of exceptional bundles on the projective plane. In particular, we will construct the following graph:

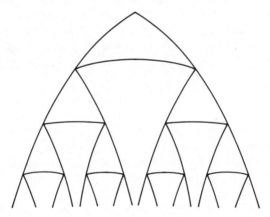

In this graph each vertex represents an exceptional pair of bundles on an elliptic curve (see def. 2); each edge a mutation of an exceptional pair; and each smooth curve a simple bundle with non-negative degree. Moreover, every exceptional pair and every simple bundle, with non-negative degree and determinant equal to a multiple of a fixed point, is represented in this graph.

In this article we will confine ourselves to the following notation: X is a nonsingular elliptic curve over the field of complex numbers, $d(E)$ is the degree of a bundle E, equal to the degree of its determinant, $r(E)$ is the rank of the bundle E, $\mathcal{E}(r,d)$ is the set of indecomposable bundles over X of rank r and degree d.

1. Properties of indecomposable bundles over an elliptic curve.

In this section we formulate and prove two lemmas on the properties of indecomposable bundles over an elliptic curve, which we will need in the proof of Atiyah's theorem.

LEMMA 1. *Let X be a nonsingular elliptic curve, E an indecomposable vector bundle over X. If $h^0(E) \neq 0$ then there exists a filtration*

$$0 = E_0 \subset E_1 \subset E_2 \subset \cdots \subset E_r = E$$

with the following properties:

1) $E_i/E_{i-1} \cong L_i$, a line bundle,

2) $h^0(L_1) \neq 0$, $h^0(L_1^* \otimes L_i) \neq 0$, where $i \geqslant 1$.

PROOF. First we note that a filtration satisfying property (1) always exists. Indeed, for any bundle E there exists $n \in \mathbf{N}$ such that $E \otimes \mathcal{O}(n)$ is generated by its global sections, where $\mathcal{O}(n)$ as usual denotes the nth tensor power of a hyperplane section. Since a filtration of $E \otimes \mathcal{O}(n)$ induces a filtration of E, we can assume that E is globally generated. Consider a nonzero homomorphism $\phi : \mathcal{O} \to E$. Since its image is a torsion-free sheaf on a curve, it is locally free, i.e. it determines a rank one subbundle of E. We will denote this subbundle by $[\phi] = E_1$. Since a quotient sheaf of a globally generated sheaf is globally generated, we can again choose a rank one subbundle L_2 of the bundle E/E_1. We define a subbundle E_2 of E by the following commutative diagram:

$$
\begin{array}{ccccccccc}
& & 0 & & 0 & & \\
& & \downarrow & & \downarrow & & \\
0 & \longrightarrow & E_1 & \longrightarrow & E_2 & \longrightarrow & L_2 & \longrightarrow & 0 \\
& & \| & & \downarrow & & \downarrow & & \\
0 & \longrightarrow & E_1 & \longrightarrow & E & \longrightarrow & E/E_1 & \longrightarrow & 0 \\
& & & & \downarrow & & \downarrow & & \\
& & & & F_1 & = & F_1 & & \\
& & & & \downarrow & & \downarrow & & \\
& & & & 0 & & 0 & &
\end{array}
$$

Continue by induction.

Next we observe that the set $\mathcal{D} = \{\deg[\phi] \mid \phi \in H^0(E),\ \phi \neq 0\}$ is bounded above. Indeed, consider some filtration of E satisfying (1) and a nonzero section $\phi \in H^0(E)$. There exists an integer $i \geqslant 1$ such that $[\phi] \subset E_i$, $[\phi] \not\subset E_{i-1}$ and ϕ defines a nonzero homomorphism $[\phi] \to L_i$, so $\deg[\phi] \leqslant \deg L_i$ and, since the set $\{\deg L_i \mid i = 1,\ldots,r\}$ is bounded, the set $\mathcal{D}$ is bounded above.

We now proceed to construct a filtration satisfying (2). Choose E_1 to be a rank one subbundle of E of maximal degree. Clearly, $h^0(L_1) \neq 0$. Assume that we have already constructed the subbundles

$$0 = E_0 \subset E_1 \subset \cdots \subset E_i \subset E$$

satisfying (2) and such that L_j has maximal degree in E/E_{j-1}. Consider the exact sequence

$$0 \longrightarrow E_i \longrightarrow E \longrightarrow E_i' \longrightarrow 0.$$

As E is indecomposable, we have $^1\langle E_i' \mid E_i \rangle \neq 0$ and, from Serre duality, there exists a nonzero homomorphism $f : E_i \to E_i'$. There is an integer j, with $1 \leqslant j \leqslant i$, such that $f(E_{j-1}) = 0$ but $f(E_j) \neq 0$. Then f induces a nonzero homomorphism $\overline{f} : L_j \to E_i'$. From the induction hypothesis $h^0(L_j) > 0$. Consequently, the composite $(\overline{f} \circ \phi)$ of $\overline{f}$ with the nonzero section ϕ turns out to be a section of E_i'. Moreover, $h^0(L_j^* \otimes [\overline{f} \circ \phi]) \neq 0$. If $[\overline{f} \circ \phi]$ has maximal degree then we choose $L_{i+1} = [\overline{f} \circ \phi]$. Otherwise, we take L_{i+1} to be to be a subbundle of E_i' of maximal degree. Then $\deg L_{i+1} \geqslant \deg[\overline{f} \circ \phi] + 1$ and, as X is an elliptic curve, $^0\langle [\overline{f} \circ \phi] \mid L_{i+1} \rangle \neq 0$, i.e. we always have $^0\langle L_1 \mid L_{i+1} \rangle \neq 0$. To complete the proof, it remains to determine E_{i+1} from the commutative diagram

$$
\begin{array}{ccccccccc}
 & & & & 0 & & 0 & & \\
 & & & & \downarrow & & \downarrow & & \\
0 & \longrightarrow & E_i & \longrightarrow & E_{i+1} & \longrightarrow & L_{i+1} & \longrightarrow & 0 \\
 & & \| & & \downarrow & & \downarrow & & \\
0 & \longrightarrow & E_i & \longrightarrow & E & \longrightarrow & E_i' & \longrightarrow & 0 \\
 & & & & \downarrow & & \downarrow & & \\
 & & & & F_i & = & F_i & & \\
 & & & & \downarrow & & \downarrow & & \\
 & & & & 0 & & 0 & &
\end{array}
$$

We denote the filtration obtained in this lemma by $(L_1, L_2, \ldots, L_r)$.

LEMMA 2. *Let $E \in \mathcal{E}(r,d)$. Then*

1) $\chi(E) = \deg E = d$.

2) If $\deg E < 0$, then $h^0(E) = 0$ and $h^1(E) = d$.

3) If $\deg E > 0$, then $h^0(E) = d$ and $h^1(E) = 0$.

4) If $d = 0$ and $h^0(E) > 0$, then $h^0(E) = 1$.

PROOF. Consider the filtration $(L_1, L_2, \ldots, L_r)$ of the bundle E. Using the additivity of the functions χ and $\deg$, it is easy to check that

$$\sum_{i=1}^{r} \chi(L_i) = \chi(E) \quad \text{and} \quad \sum_{i=1}^{r} \deg L_i = \deg E.$$

Now, for line bundles, we have the Riemann-Roch formula $\chi(L_i) = \deg L_i$, so the first part of the lemma is clear.

Next, under the hypothesis of part (2), assume that $h^0(E) > 0$. Then we may require the filtration to satisfy $h^0(L_i) \neq 0$. But, since $\deg L_i \geqslant 0$, the degree of E is also non-negative, which contradicts the hypothesis. The third statement of the lemma follows easily from the second and Serre duality.

Now let the degree of E be equal to zero and $h^0(E) > 0$. Then the sum of the nonnegative degrees of the L_i (the elements of the filtration of the bundle E) is equal to zero and thus $\deg L_i = 0$ for all i. But all the L_i have sections, which is possible only when they are trivial.

We assume that E_i is the smallest element of the filtration satisfying the condition $h^0(E_i) \geqslant 2$, then $E_i \cong \mathcal{O} \oplus E_{i-1}$. Indeed, E_i is included in the exact sequence

$$0 \longrightarrow E_{i-1} \longrightarrow E_i \longrightarrow \mathcal{O} \longrightarrow 0. \tag{1}$$

and the coboundary homomorphism $\delta : H^0(\mathcal{O}) \to H^1(E_{i-1})$ is zero. Therefore sequence (1) splits. The space $H^k(E_i)$ includes the distinguished line $H^k(E_{i-1})$ $(k = 0, 1)$. The direct summand $\mathcal{O}$ determines another line in $H^k(E_i)$ which is different from $H^k(E_{i-1})$. It is easy to see that any line $\ell \in H^k(E_i)$ different from $H^k(E_{i-1})$ determines a decomposition $E_i = E_{i-1} \oplus \mathcal{O}$ such that $H^k(\mathcal{O}) = \ell$.

For the next term in the filtration the space of global sections is not smaller than that of E_i. Let us assume that $h^0(E_{i+1}) = h^0(E_i)$. From the universal property there exists a commutative diagram

$$
\begin{array}{ccccccc}
0 & \longrightarrow & E_i & \longrightarrow & E'_{i+1} & \longrightarrow & H^1(E_i) \otimes \mathcal{O} & \longrightarrow & 0 \\
 & & \| & & \uparrow & & \uparrow i & & \\
0 & \longrightarrow & E_i & \longrightarrow & E_{i+1} & \longrightarrow & \mathcal{O} & \longrightarrow & 0
\end{array}
\tag{2}
$$

where the first row is the universal extension. We assume that $\text{im}(i)$ does not coincide with $H^1(E_{i-1})$. Then the line $\ell = H^1(\text{im}(i))$ gives a decomposition of diagram (2):

$$0 \longrightarrow E_{i-1} \oplus \mathcal{O} \longrightarrow \tilde{E}_i \oplus F_2 \longrightarrow H^1(E_{i-1}) \otimes \mathcal{O} \oplus \ell \otimes \mathcal{O} \longrightarrow 0$$

$$\| \qquad\qquad \uparrow \qquad\qquad\qquad \uparrow$$

$$0 \longrightarrow E_{i-1} \oplus \mathcal{O} \longrightarrow E_{i+1} \xrightarrow{\hspace{3cm}} \mathcal{O} \longrightarrow 0$$

where F_2 is the universal extension of $\mathcal{O}$ by $\mathcal{O}$ and $\tilde{E}_i$ is the universal extension of $\mathcal{O}$ by E_{i-1}. Therefore $E_{i+1} \cong E_{i-1} \oplus F_2$. This decomposition determines a line $\ell = H^k(F_2)$ in the space $H^k(E_{i+1})$ ($k = 0, 1$), which does not coincide with the distinguished line $H^k(E_{i-1})$. Conversely, any line $\ell \neq H^k(E_{i-1})$ in the space $H^k(E_{i+1})$ determines a decomposition into a direct sum $E_{i+1} \cong E_{i-1} \oplus F_2$. Indeed, let $\ell \subset H^k(E_{i+1})$ but $\ell \not\subset H^k(E_{i-1})$. If we consider it as a line in $H^k(E_i) \cong H^k(E_{i+1})$, we get a decomposition $E_i = E_{i-1} \oplus \mathcal{O}$, where $H^k(\mathcal{O}) = \ell$. In the space $H^1(E_i)$ there is a line ℓ' corresponding to the line ℓ, which coincides with ℓ if, from the very beginning, $k = 1$. Then E_{i+1} can be included in the sequence

$$0 \longrightarrow E_{i-1} \oplus \mathcal{O} \longrightarrow E_{i+1} \longrightarrow \ell' \otimes \mathcal{O} \longrightarrow 0,$$

i.e. $E_{i+1} \cong E_{i-1} \oplus F_2$ and $H^k(F_2) = \ell$.

If $\text{im}(i)$ in diagram (2) coincides with $H^1(E_{i-1})$ then $E_{i+1} \cong \tilde{E}_i \oplus \mathcal{O}$. In the space $H^k(E_{i+1})$ the line $H^k(\tilde{E}_i) = H^k(E_i)$ is distinguished and any other line ℓ determines a decomposition $E_{i+1} \cong \tilde{E}_i \oplus \mathcal{O}$, such that $H^k(\mathcal{O}) = \ell$.

If $h^0(E_{i+1}) > h^0(E_i)$ then the sequence

$$0 \longrightarrow E_i \longrightarrow E_{i+1} \longrightarrow \mathcal{O} \longrightarrow 0$$

splits and $E_{i+1} \cong E_i \oplus \mathcal{O}$. Such a decomposition is not unique but corresponds to a line in $H^k(E_{i+1}) \backslash H^k(E_i)$.

The symbol F_s will denote the universal extension of $\mathcal{O}$ by F_{s-1}, for $s = 3, 4, \ldots$; the symbol $\tilde{E}_s$ will denote the universal extension of $\mathcal{O}$ by $\tilde{E}_{s-1}$, for $s = i, i+1, \ldots, r$.

Suppose that we have established the possibility of a direct sum decomposition

$$E_{r-1} \cong \tilde{F}_1 \oplus \cdots \oplus \tilde{F}_m$$

where $\tilde{F}_s$ belongs to the set $\{\mathcal{O}, F_k\,(k = 2, 3, \ldots), \tilde{E}_k\,(k = i, i+1, \ldots, r)\}$. There is a natural filtration of E_{r-1}

$$G_1 \subset G_2 \subset \cdots \subset G_m = E_{r-1}, \quad \text{with} \quad G_1 = \tilde{F}_1, \ldots, G_m = G_{m-1} \oplus \tilde{F}_m,$$

which induces a filtration of $H^k(E_{r-1})$

$$V_1 \subset V_2 \subset \cdots \subset V_m = H^k(E_{r-1}), \quad \text{with} \quad V_s = H^k(G_s).$$

Any line ℓ in V_s which is not in V_{s-1} gives a decomposition $G_s = G_{s-1} \oplus \tilde{F}_s$, such that $H^k(\tilde{F}_s) = \ell$.

If $h^0(E_r) > h^0(E_{r-1})$ then, as above, it is easy to show that E_r is decomposable using the fact that the corresponding sequence splits. Now assume that $h^0(E_r) = h^0(E_{r-1})$. Consider the commutative diagram

$$
\begin{array}{ccccccccc}
0 & \longrightarrow & E_{r-1} & \longrightarrow & \tilde{E} & \longrightarrow & H^1(E_{r-1}) \otimes \mathcal{O} & \longrightarrow & 0 \\
& & \| & & \uparrow & & \uparrow{\scriptstyle i} & & \\
0 & \longrightarrow & E_{r-1} & \longrightarrow & E & \longrightarrow & \mathcal{O} & \longrightarrow & 0
\end{array}
$$

There exists a number s such that $\mathrm{im}(i) \subset G_s$ but $\mathrm{im}(i) \not\subset G_{s-1}$. Then $\ell = H^1(\mathrm{im}(i))$ gives a decomposition

$$E_{r-1} \cong G_{s-1} \oplus \tilde{F}_s \oplus \cdots \oplus \tilde{F}_m,$$

and $H^1(\tilde{F}_{s+1}) = \ell$. Therefore E can be included in the exact sequence

$$0 \longrightarrow E_{r-1} \longrightarrow E \longrightarrow \ell \otimes \mathcal{O} \longrightarrow 0$$

and $E \cong G_{s-1} \oplus \hat{F}_s \oplus \tilde{F}_{s+1} \oplus \cdots \oplus \tilde{F}_m$, where $\hat{F}_s$ is the universal extension of $\mathcal{O}$ by $\tilde{F}_s$, i.e. we get a decomposition of E into a direct sum, which contradicts the assumption.

2. Mutations of indecomposable and simple bundles.

In this section we will define a simple pair of bundles on an elliptic curve. We will prove that mutations preserve simplicity of pairs and we will also explore the possible existence of mutations of indecomposable bundles.

DEFINITION 1. A *simple pair* is an ordered pair of simple bundles A and B, for which ${}^i\langle A \mid B \rangle \neq 0$ for only one value of i.

LEMMA 3. Let E_1 be a simple bundle.

1) If the sequence

$$0 \longrightarrow L \xrightarrow{\;i\;} V \otimes E_1 \xrightarrow{\;p\;} E_2 \longrightarrow 0 \tag{3}$$

is exact then the following conditions are equivalent:

a) (E_1, E_2) is a simple pair, $V \cong {}^0\langle E_1 \mid E_2 \rangle$,

b) (L, E_1) is a simple pair, $V \cong {}^0\langle L \mid E_1 \rangle^*$.

2) If the sequence

$$0 \longrightarrow V \otimes E_1 \xrightarrow{i} E_2 \xrightarrow{p} L \longrightarrow 0 \qquad (4)$$

is exact then the following conditions are equivalent:

a) (E_1, E_2) *is a simple pair,* $\qquad V \cong {}^0\langle E_1 \mid E_2 \rangle$,

b) (L, E_1) *is a simple pair,* $\qquad V \cong {}^1\langle L \mid E_1 \rangle^*$.

3) If the sequence

$$0 \longrightarrow E_2 \xrightarrow{i} L \xrightarrow{p} V \otimes E_1 \longrightarrow 0 \qquad (5)$$

is exact then the following conditions are equivalent:

a) (E_1, E_2) *is a simple pair,* $\qquad V \cong {}^1\langle E_1 \mid E_2 \rangle$,

b) (L, E_1) *is a simple pair,* $\qquad V \cong {}^0\langle L \mid E_1 \rangle^*$.

PROOF. Recall that for every exact sequence

$$0 \longrightarrow A \longrightarrow B \longrightarrow C \longrightarrow 0$$

there is a corresponding commutative diagram:

$$
\begin{array}{ccccccccccc}
& 0 & & 0 & & 0 & & \vdots & & \vdots & & \vdots \\
& \downarrow & & \downarrow & & \downarrow & & \downarrow & & \downarrow & & \downarrow \\
0 \to & {}^0\langle C \mid A \rangle & \to & {}^0\langle C \mid B \rangle & \to & {}^0\langle C \mid C \rangle & \to & {}^1\langle C \mid A \rangle & \to & {}^1\langle C \mid B \rangle & \to & {}^1\langle C \mid C \rangle & \to 0 \\
& \downarrow & & \downarrow & & \downarrow & & \downarrow & & \downarrow & & \downarrow \\
0 \to & {}^0\langle B \mid A \rangle & \to & {}^0\langle B \mid B \rangle & \to & {}^0\langle B \mid C \rangle & \to & {}^1\langle B \mid A \rangle & \to & {}^1\langle B \mid B \rangle & \to & {}^1\langle B \mid C \rangle & \to 0 \\
& \downarrow & & \downarrow & & \downarrow & & \downarrow & & \downarrow & & \downarrow \\
0 \to & {}^0\langle A \mid A \rangle & \to & {}^0\langle A \mid B \rangle & \to & {}^0\langle A \mid C \rangle & \to & {}^1\langle A \mid A \rangle & \to & {}^1\langle A \mid B \rangle & \to & {}^1\langle A \mid C \rangle & \to 0 \\
& \downarrow & & \downarrow & & \downarrow & & \downarrow & & \downarrow & & \downarrow \\
\cdots \to & {}^1\langle C \mid A \rangle & \to & {}^1\langle C \mid B \rangle & \to & {}^1\langle C \mid C \rangle & \to & 0 & & 0 & & 0 \\
& \downarrow & & \downarrow & & \downarrow \\
\cdots \to & {}^1\langle B \mid A \rangle & \to & {}^1\langle B \mid B \rangle & \to & {}^1\langle B \mid C \rangle & \to & 0 \\
& \downarrow & & \downarrow & & \downarrow \\
\cdots \to & {}^1\langle A \mid A \rangle & \to & {}^1\langle A \mid B \rangle & \to & {}^1\langle A \mid C \rangle & \to & 0 \\
& \downarrow & & \downarrow & & \downarrow \\
& 0 & & 0 & & 0
\end{array}
$$

Suppose that property (1a) is satisfied. From the sequence

$$0 \longrightarrow {}^0\langle E_1 \mid L \rangle \longrightarrow V \otimes {}^0\langle E_1 \mid E_1 \rangle \xrightarrow{\sim} {}^0\langle E_1 \mid E_2 \rangle \longrightarrow \cdots$$

it follows that ${}^0\langle E_1 \mid L \rangle \cong {}^1\langle L \mid E_1 \rangle^* = 0$. In addition ${}^0\langle E_2 \mid E_1 \rangle \cong {}^1\langle E_1 \mid E_2 \rangle^* = 0$, because (E_1, E_2) is a simple pair and ${}^0\langle E_1 \mid E_2 \rangle \neq 0$. If the condition (1b) is satisfied, then, in the sequence (3), i is a canonical map and, in the sequence

$$0 \longrightarrow {}^0\langle E_2 \mid E_1 \rangle \longrightarrow V^* \otimes {}^0\langle E_1 \mid E_1 \rangle \xrightarrow{\alpha} {}^0\langle L \mid E_1 \rangle \longrightarrow \cdots,$$

α is an isomorphism. Therefore $^0\langle E_2 \mid E_1 \rangle \cong {}^1\langle E_1 \mid E_2 \rangle^* = 0$. From the condition (1)(b) it follows that $^0\langle E_1 \mid L \rangle \cong {}^1\langle L \mid E_1 \rangle^* = 0$.

We now introduce the following notation:

$$V_1 = {}^0\langle E_1 \mid E_2 \rangle, \quad V_2 = {}^0\langle L \mid E_1 \rangle^*, \quad W = {}^0\langle L \mid E_2 \rangle.$$

Using this notation we write down part of the diagram corresponding to the exact sequence (3), taking into account the equalities we have obtained.

$$
\begin{array}{c}
 0 \\
 \downarrow \\
0 \longrightarrow V \otimes V^* \otimes {}^0\langle E_1 \mid E_1 \rangle \xrightarrow{\;\alpha\;} V^* \otimes V_1 \longrightarrow V^* \otimes V_2 \xrightarrow{\;\varepsilon\;} V \otimes V^* \otimes \mathbb{C} \longrightarrow 0 \\
 \downarrow{\scriptstyle \beta} \downarrow{\scriptstyle \phi} \downarrow \\
 V \otimes V_2^* \xrightarrow{\;\psi\;} W \longrightarrow {}^1\langle L \mid L \rangle \longrightarrow 0 \\
 \downarrow \downarrow \\
 V \otimes V_1^* \longrightarrow {}^1\langle E_2 \mid E_2 \rangle \\
 \downarrow{\scriptstyle \gamma} \downarrow \\
V \otimes V^* \otimes {}^1\langle E_1 \mid E_1 \rangle 0 \\
 \downarrow \\
 0
\end{array}
\tag{6}
$$

If $V \cong V_1$ then by considering dimensions we see that $\alpha, \gamma, \varepsilon$ and β are isomorphisms and diagram (6) reduces to the diagram

$$
\begin{array}{ccccccc}
V^* \otimes V_1 & \xrightarrow{\;\phi\;} & W & \longrightarrow & {}^1\langle E_2 \mid E_2 \rangle & \longrightarrow & 0 \\
\wr\uparrow & & \| & & & & \\
V^* \otimes V_2 & \xrightarrow{\;\psi\;} & W & \longrightarrow & {}^1\langle L \mid L \rangle & \longrightarrow & 0.
\end{array}
\tag{7}
$$

From this we get that $^1\langle L \mid L \rangle \cong {}^1\langle E_2 \mid E_2 \rangle$ and $V_2 \cong V$.

If $V \cong V_2$ then considering dimensions we see that $\beta, \varepsilon, \gamma$ and α are isomorphisms and diagram (6) again reduces to diagram (7), which proves the first part of the lemma.

Proof of the remaining parts reduces to considering diagram (6) with the following notation: $V_1 = {}^0\langle E_1 \mid E_2 \rangle$, $V_2 = {}^1\langle L \mid E_1 \rangle^*$ in the first case; $V_1 = {}^1\langle E_1 \mid E_2 \rangle$, $V_2 = {}^0\langle L \mid E_1 \rangle^*$ in the second case. In both cases $W = {}^1\langle L \mid E_2 \rangle$.

REMARK From the proof of the lemma it follows that, if $^0\langle E_1 \mid E_2 \rangle \neq 0$, $^1\langle E_1 \mid E_2 \rangle = 0$ and E_1 is a simple bundle, then, when E_2 is not simple, L is not simple. Conversely, if E_1 is simple, $^1\langle L \mid E_1 \rangle \neq 0$ and $^0\langle L \mid E_1 \rangle = 0$, then, when L is not simple, E_2 is not simple.

LEMMA 4. *Let $E \in \mathcal{E}(r,d)$, with $d > 0$ or $d = 0$ and $H^0(E) \neq 0$.*

1) *For the pair $(E, \mathcal{O})$ there is a right mutation of type "ext"*

$$0 \longrightarrow {}^1\langle E \mid \mathcal{O} \rangle^* \otimes \mathcal{O} \xrightarrow{\ i\ } E' \longrightarrow E \longrightarrow 0, \tag{8}$$

where E' is an indecomposable bundle in $\mathcal{E}(r + d, d)$, if $d \neq 0$, or in $\mathcal{E}(r + 1, 0)$, if $d = 0$. There is a canonical isomorphism $^0\langle \mathcal{O} \mid E' \rangle^* \cong {}^1\langle E \mid \mathcal{O} \rangle$ and the pair $(\mathcal{O}, E')$ can be left mutated to give E.

2) *If $0 \leqslant d < r$, then for the pair $(\mathcal{O}, E)$ there is a left mutation*

$$0 \longrightarrow H^0(E) \otimes \mathcal{O} \longrightarrow E \longrightarrow E' \longrightarrow 0. \tag{9}$$

The bundle E' is contained in $\mathcal{E}(r - d, d)$, if $d \neq 0$, or in $\mathcal{E}(r - 1, 0)$, if $d = 0$. There is a canonical isomorphism $^1\langle E' \mid \mathcal{O} \rangle^* \cong H^0(E)$ and the result of a right mutation on the pair $(E', \mathcal{O})$ is E.

PROOF.

1) From Lemma 2 it follows that

$$\dim {}^1\langle E \mid \mathcal{O} \rangle = \begin{cases} d & \text{if } d > 0 \\ 1 & \text{if } d = 0, \ h^0(E) \neq 0 \end{cases}$$

Therefore sequence (8) is nontrivial. Applying the functor $\mid \mathcal{O} \rangle$ to it, we obtain

$$\cdots {}^1\langle E \mid \mathcal{O} \rangle \xrightarrow{\ \delta\ } {}^1\langle E \mid \mathcal{O} \rangle \longrightarrow {}^1\langle E' \mid \mathcal{O} \rangle \longrightarrow {}^1\langle E \mid \mathcal{O} \rangle \longrightarrow 0,$$

where δ is an isomorphism, because the extension is universal, and so $^0\langle \mathcal{O} \mid E' \rangle^* \cong {}^1\langle E' \mid \mathcal{O} \rangle \cong {}^1\langle E \mid \mathcal{O} \rangle$. If in sequence (8) we change the space $^1\langle E \mid \mathcal{O} \rangle^*$ to $^0\langle \mathcal{O} \mid E' \rangle$, then the map i is still canonical, showing that the pair $(\mathcal{O}, E')$ can be mutated.

If we assume that $E' = F \oplus G$, then $H^0(E') = H^0(F) \oplus H^0(G)$ and the sequence (8) can be rewritten in the form

$$0 \longrightarrow \left(H^0(F) \otimes \mathcal{O} \right) \oplus \left(H^0(G) \otimes \mathcal{O} \right) \xrightarrow{\ can \oplus can\ } F \oplus G \longrightarrow E \longrightarrow 0.$$

This means that E is decomposable, which contradicts the assumption.

2) Consider the filtration $(L_1, \ldots, L_r)$ of the bundle E as constructed in Lemma 1. If $\deg L_1 \geqslant 1$, then $\deg E = \sum \deg L_i \geqslant r$. Therefore $\deg L_1 = 0$ and, since $h^0(L_1) \neq 0$, $L_1 \cong \mathcal{O}$. But L_1 is a subbundle of E which has maximal degree among one-dimensional subbundles of E. Therefore any holomorphic section of E has degree zero and the

canonical map $H^0(E) \otimes \mathcal{O} \to E$ is an injection. Consider the long exact sequence corresponding to the sequence (9)

$$0 \to H^0(E) \xrightarrow{i_*} H^0(E) \to H^0(E') \xrightarrow{\delta} H^0(E) \to H^1(E) \xrightarrow{\alpha} H^1(E') \to 0.$$

It follows from the construction that i_* is an isomorphism. Since E has non-negative degree, $h^1(E) = 0$ or $h^1(E) = 1$. If $d > 0$ then $h^1(E) = 0$ and δ is an isomorphism. If $d = 0$ and $h^1(E') = 0$ then the Riemann-Roch theorem implies the triviality of the space $^0\langle \mathcal{O} \mid E' \rangle \cong {}^1\langle E' \mid \mathcal{O} \rangle^*$, i.e. the sequence (9) splits, which contradicts the indecomposability of E. Therefore $h^1(E') = 1$ and α is an isomorphism. Consequently δ is also an isomorphism, i.e. $^1\langle E' \mid \mathcal{O} \rangle^* \cong {}^0\langle \mathcal{O} \mid E' \rangle \cong H^0(E)$, and one can assume that the sequence (9) represents a right mutation of the pair $(E', \mathcal{O})$.

Now assume that E' is represented as a direct sum of nonzero bundles F' and G'. We let F and G denote the results of applying right mutations of type "ext" to the pairs $(F', \mathcal{O})$ and $(G', \mathcal{O})$ respectively. Then $E \cong F \oplus G$, which contradicts the indecomposability of E.

3.   The Theorems of Atiyah.

THEOREM 1.

1) *For every $r > 0$ there exists a bundle $F_r \in \mathcal{E}(r, 0)$, unique up to isomorphism, such that $h^0(F_r) > 0$. Furthermore, $F_1 \cong \mathcal{O}$ and each F_r can be obtained as the universal extension of F_{r-1} by $\mathcal{O}$.*

2) *Every bundle $E \in \mathcal{E}(r, 0)$ can be obtained from F_r as a tensor product with a root of its determinant, i.e. $E \cong F_r \otimes L$, where $L^r \cong \det E$.*

PROOF.

1) The trivial bundle is the only bundle in $\mathcal{E}(1, 0)$ which has nonzero global sections. Since $h^1(\mathcal{O}_X) = 1$, there exists a universal extension

$$0 \to \mathcal{O}_X \to F_2 \to H^1(\mathcal{O}_X) \otimes \mathcal{O}_X \to 0.$$

Moreover, it follows from the long exact cohomology sequence that $H^i(F_2) \cong H^i(\mathcal{O}_X)$. Again there is a universal extension

$$0 \to \mathcal{O}_X \to F_3 \to {}^1\langle F_2 \mid \mathcal{O}_X \rangle \otimes F_2 \to 0,$$

and moreover $H^0(F_3) \cong H^0(\mathcal{O}_X)$ and $h^1(F_3) = 1$; and so on.

This argument proves the existence of bundles in $\mathcal{E}(r,0)$ which have nonzero sections. For uniqueness, let us assume that it has been demonstrated for $\mathcal{E}(r',0)$ with $r' < r$ and let $E \in \mathcal{E}(r,0)$ with $h^0(E) > 0$. From Lemma 2 we know that $h^0(E) = 1$ and, by Lemma 4, there exists an exact sequence

$$0 \to H^0(E) \otimes \mathcal{O}_X \to E \to E' \to 0,$$

where $E' \in \mathcal{E}(r-1,0), h^0(E') = h^1(E') = 1$ and the given sequence is a universal extension of E' by $\mathcal{O}_X$. By the induction hypothesis, $E' \cong F_{r-1}$. Then the uniqueness of the universal extension implies that $E \cong F_r$.

2) Consider the filtration $(L_1, L_2, \ldots, L_r)$ of $E \in \mathcal{E}(r,0)$. The bundle $E \otimes A$, for $A \in \mathcal{E}(1,1)$, is contained in $\mathcal{E}(r,r)$. If $\deg(L_1 \otimes A) = 0$, then $E \otimes A$ is trivial. But $\deg(L_1 \otimes A) \leqslant 1$, because $r\big(\deg(L_1 \otimes A)\big) \leqslant r$. Therefore $\deg L_1 = 0$. The bundle $L_1^* \otimes E$ is contained in the set $\mathcal{E}(r,0)$ and is isomorphic to F_r, because $h^0(L_1^* \otimes E) \neq 0$ $(\mathcal{O}_X \subset E)$. Consequently $E \cong F_r \otimes L_1$ and $L_1^r \cong \det E$, because $\det F_r \cong \mathcal{O}_X$.

THEOREM 2. *Let A be a fixed line bundle on X of degree one. Then A determines a one-one correspondence $\alpha_{r,d} : \mathcal{E}(h,0) \to \mathcal{E}(r,d)$, where h is the highest common factor of r and d. The map $\alpha_{r,d}$ is defined inductively.*

1) $\alpha_{r,0} = id.$

2) $\alpha_{r,d+r}(E) \cong \alpha_{r,d}(E) \otimes A.$

3) If $0 < d < r$, then $\alpha_{r,d}(E)$ is determined by the following sequence:

$$0 \to {}^1\langle \alpha_{r-d,d}(E) \mid \mathcal{O}_X \rangle \otimes \mathcal{O}_X \to \alpha_{r,d}(E) \to \alpha_{r-d,d}(E) \to 0.$$

PROOF. The proof of this theorem is easily obtained by applying Lemma 4, so we will omit it.

THEOREM 3. *An indecomposable bundle on a non-singular elliptic curve is simple if and only if the highest common factor of its rank and degree is equal to one.*

PROOF. It follows from Theorem 1 that every indecomposable bundle E of rank r and degree zero is isomorphic to $F_r \otimes L$, for a line bundle L. Consequently ${}^0\langle E \mid E \rangle \cong {}^0\langle F_r \mid F_r \rangle$. The bundle F_r has nonzero sections and, from the Riemann-Roch theorem, F_r^* also has nonzero sections. Therefore the bundle F_r is simple only if $r = 1$ (see [2] p.75-76).

It follows from Theorem 2 that every bundle $E \in \mathcal{E}(r,d)$ is equal to $\alpha_{r,d}(E')$, for some $E' \in \mathcal{E}\big((r,d),0\big)$. The map $\alpha_{r,d}$ is a composed of the following maps:

$$t : \mathcal{E}(r,d) \to \mathcal{E}(r,d+r), \qquad t(E) \cong E \otimes A;$$

$$ext : \mathcal{E}(r,d) \to \mathcal{E}(r+d,d)$$

where $0 < d < r$ and $ext(E)$ is the universal extension of E by $\mathcal{O}_X$. It is evident that ${}^0\langle t(E) \mid t(E) \rangle \cong {}^0\langle E \mid E \rangle$, and from the remark after lemma 3, it follows that ${}^0\langle ext(E) \mid ext(E) \rangle \cong {}^0\langle E \mid E \rangle$. Therefore $\alpha_{r,d}$ maps simple bundles to simple bundles and nonsimple ones to nonsimple ones.

4. The Coprimality Equation and the Classification of Simple Bundles.

According to the results of Atiyah, indecomposable bundles on a nonsingular elliptic curve are simple if and only if their rank and degree are coprime. In this section we will describe the construction of simple bundles with determinant a multiple of $\mathcal{O}(p)$, where p is a fixed point of the curve. This problem is closely related to the equation

$$x_1 x_4 - x_2 x_3 = 1. \tag{10}$$

Indeed, if (a_1, a_2, a_3, a_4) is a solution of this equation with $a_1 > 0$, there exists a simple bundle of rank a_1, with determinant equal to $\mathcal{O}(a_2 p)$. Conversely, if E is a simple bundle of rank r and degree d, there exist integers r', d', such that $rd' - r'd = 1$.

Just as the Markov process describes solutions of the Markov equations, there exists a process describing the solutions of equation (10).

There are two functions defined on the set of integer solutions of equation (10):

$$e_1(x_1, x_2, x_3, x_4) = (x_1, x_2, x_1 + x_3, x_2 + x_4)$$
$$e_2(x_1, x_2, x_3, x_4) = (x_1 + x_3, x_2 + x_4, x_3, x_4).$$

It is easy to check that these do act on the set of solutions of equation (10). As usual, we will call these functions mutations of solutions. The monoid generated by maps e_1 and e_2 will be denoted by G. We shall denote by M the set of nonnegative integer solutions of equation (10).

LEMMA 5. Let $x = (x_1, x_2, x_3, x_4) \in M$. If $x_1 > x_3$ and $x_4 > x_2$ then $x = (1, 0, 0, 1)$.

PROOF. The assumptions of the lemma imply that the following chain of inequalities is satisfied:

$$1 = x_1 x_4 - x_2 x_3 \geqslant x_3 x_4 + x_4 - x_3 x_4 + x_3 = x_4 + x_3.$$

The coordinate x_4 cannot be equal to zero, because this would contradict the inequalities $x_4 > x_2 \geqslant 0$. Therefore $x_4 = 1$, $x_3 = 0$, $x_2 = 0$ and $x_1 = 1$.

COROLLARY. The following mutations are defined on the set M:

$$h_1(x) = \begin{cases} (x_1 - x_3, x_2 - x_4, x_1, x_2) & \text{if } x_1 \geqslant x_3 \text{ and } x_2 \geqslant x_4 \\ (x_1, x_2, x_3 - x_1, x_4 - x_2) & \text{if } x_1 \leqslant x_3 \text{ and } x_2 \leqslant x_4 \end{cases}$$

$$h_2(x) = \begin{cases} (x_3, x_4, x_3 - x_1, x_4 - x_2) & \text{if } x_1 \leqslant x_3 \text{ and } x_2 \leqslant x_4 \\ (x_1 - x_3, x_2 - x_4, x_3, x_4) & \text{if } x_1 \geqslant x_3 \text{ and } x_2 \geqslant x_4 \end{cases}$$

By some easy calculations, we deduce the following properties of the mutations:

LEMMA 6.

1) $h_1 e_1 = id$, $h_2 e_2 = id$, $h_2 e_1 = e_2$ and $h_1 e_2 = e_1$.

2) If $x_1 \leqslant x_3$ and $x_2 \leqslant x_4$, then $h_1 h_2 = id$ and $e_1 h_1 = id$.

3) If $x_1 \geqslant x_3$ and $x_2 \geqslant x_4$, then $h_2 h_1 = id$ and $e_2 h_2 = id$.

PROPOSITION 7. *Let the quadruple of nonnegative integers $x = (x_1, x_2, x_3, x_4)$ satisfy equation (10). Then there exists $e \in G$ such that $e(1, 0, 0, 1) = x$.*

PROOF. By induction with respect to $|x| = x_1 + x_2 + x_3 + x_4$ (the *norm* of x). If $|x| = 2$ then either $x_j = 2$ and $x_i = 0$ for $i \neq j$, which is not possible, or $x = (1, 0, 0, 1)$ and the group element we need is just the identity.

Assuming that the proposition is proved for all $x \in M$ with norm less than N, consider $x = (x_1, x_2, x_3, x_4) \in M$ with $|x| = N$. Lemma 5 implies that there are two possibilities: (a) $x_1 \geqslant x_3$, $x_2 \geqslant x_4$ or (b) $x_1 \leqslant x_3$, $x_2 \leqslant x_4$. In case (a) we can apply the mutation h_2 to x to get a vector in M with $|h_2(x)| < N$. Lemma 6 gives $e_2 h_2(x) = x$, but the induction hypothesis implies that there is an element e in the group G such that $e(1, 0, 0, 1) = h_2(x)$. Hence $e_2 e(1, 0, 0, 1) = x$. In case (b) we can apply the mutation h_1. Again, there exists $e \in G$ such that $e(1, 0, 0, 1) = h_1(x)$ and so $e_1 e(1, 0, 0, 1) = x$. This completes the proof of the proposition.

The process of generating solutions to (10) can be illustrated in the following picture:

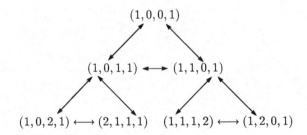

In this section, the bundles we shall be considering will all have determinant equal to a multiple of a fixed point on the elliptic curve X. Once we have a description of the solutions of equation (10), it is easy to give an algorithm for constructing simple bundles.

DEFINITION 2. An ordered pair of simple bundles (A, B) is called *exceptional* if it is simple (see Definition 1 in Section 2) and the nonzero space $^i\langle A \mid B \rangle$ is one-dimensional.

Serre Duality shows that (A, B) is exceptional if and only if (B, A) is exceptional. Therefore it is convenient to represent exceptional pairs as a quiver

$$A \rightleftarrows B$$

Consider the skyscraper sheaf $\mathcal{O}_p$ at the point p. Since $^0\langle \mathcal{O}_X \mid \mathcal{O}_p \rangle \cong \mathbb{C}$ and $^0\langle \mathcal{O}_p \mid \mathcal{O}_X \rangle = 0$, it is convenient to also regard the pair of sheaves $(\mathcal{O}_X, \mathcal{O}_p)$ as exceptional.

To an exceptional pair (A, B) satisfying the condition $^0\langle A \mid B \rangle \neq 0$ we will associate an integer vector $v(A, B) = (r(A), d(A), r(B), d(B))$. We note first that $v(A, B)$ is a solution of (10) because

$$\chi(A^* \otimes B) - r(A)d(B) - r(B)d(A) = 1.$$

We will denote by $\mathcal{O}(k)$ $(k \in \mathbf{Z})$ the line bundle on the elliptic curve X associated to the divisor kp. Now let us consider the exceptional pair $(\mathcal{O}_X, \mathcal{O}_p)$. The corresponding vector is $(1, 0, 0, 1)$. The mutations of type "ext" applied to $(\mathcal{O}_X, \mathcal{O}_p)$ give the pairs $(\mathcal{O}(1), \mathcal{O}_p)$ (left mutation) and $(\mathcal{O}, \mathcal{O}(1))$ (right mutation). On these pairs mutations of type "hom" are possible. The left mutation of $(\mathcal{O}(1), \mathcal{O}_p)$ is $(\mathcal{O}, \mathcal{O}(1))$ and the right mutation of $(\mathcal{O}, \mathcal{O}(1))$ is $(\mathcal{O}(1), \mathcal{O}_p)$. The correspondence between the described pairs and their mutations is illustrated in the picture

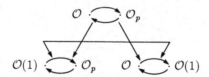

It is easy to see that the process of generating exceptional pairs can be continued indefinitely. Moreover, it is analogous to the process of generating solutions to equation (10).

If we represent exceptional pairs by vertices and mutations by arcs, then we get the following picture:

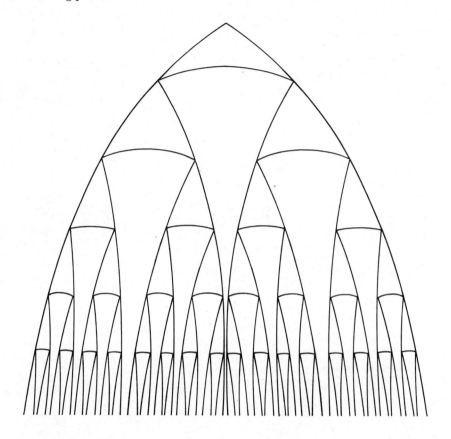

In this picture, four arcs meet at every vertex except the top one. An arc going left (resp. right) and down represents a left (resp. right) mutation of type "*ext*"; one going across to the right (resp. left) represents a left (resp. right) mutation of type "*hom*". One then easily sees that the smooth curves in this picture correspond to exceptional bundles. More precisely, if a curve starts at the pair (A,B) and goes down to the left (resp. right), then it goes through the pairs of the form $(L_B^n A, B)$ (resp. $(A, R_A^n B)$).

The following theorem is obvious from Proposition 7.

THEOREM 4. *Every simple bundle with determinant a multiple of a point p on a nonsingular elliptic curve and every exceptional pair composed of such bundles is obtained by mutations from the pair $(\mathcal{O}_X, \mathcal{O}_p)$.*

References

[1] ATIYAH, M.F., Vector Bundles Over an Elliptic Curve, *Proc. Lond. Math. Soc*, **VII** (1957) 414–452.

[2] OKONEK, K., SCHNEIDER, M. & SPINDLER, H., *Vector Bundles on Complex Projective Space*, Birkhäuser (1980).

3. Computing Invariants for Exceptional Bundles on a Quadric

S. K. Zube   D. Yu. Nogin

Introduction.

The study of exceptional bundles on $\mathbf{P}^2$ shows how important they are in relation to other bundles. So it is natural to consider this notion for other varieties and to try to describe and develop methods for analysing exceptional bundles. For many reasons the most convenient variety to look at, other than $\mathbf{P}^2$, is the quadric $\mathbf{P}^1 \times \mathbf{P}^1$. The most important properties of exceptional bundles, namely stability and their characterisation by rank and first Chern class, are preserved because they are homogeneous and because the anticanonical class is ample.

This paper consists of two parts. In the first part the definitions and basic properties of exceptional bundles on $\mathbf{P}^1 \times \mathbf{P}^1$ are given. In the second part we study the restrictions of these bundles to lines and elliptic curves. This will allow us to exhibit some canonical short exact sequences for an arbitrary exceptional bundle.

The authors are very grateful to A. N. Rudakov and A. N. Tyurin for their advice and interest which were indispensable in the writing of this paper.

1.   Basic Properties of Exceptional Bundles on $\mathbf{P}^1 \times \mathbf{P}^1$.

A bundle E on $\mathbf{P}^1 \times \mathbf{P}^1$ is called *exceptional* if

$$\mathrm{Hom}(E, E) = \mathbf{C} \qquad \text{and} \qquad \mathrm{Ext}^1(E, E) = 0$$

i.e. the bundle is simple and rigid. From simplicity and the fact that the anticanonical class on a quadric is effective it follows that $\mathrm{Ext}^2(E, E) = 0$. Indeed, from Serre duality $\left(\mathrm{Ext}^2(E, E)\right)^* = \mathrm{Ext}^0(E(-K), E)$ and if we suppose that $\mathrm{Ext}^2(E, E) \neq 0$ then there is a nontrivial homomorphism $h : E(-K) \to E$. Then h can be regarded as an endomorphism of E vanishing along the curve $-K$, so that h is not a homothety in which case $\dim \mathrm{Hom}(E, E) \geqslant 2$, which contradicts the assumption that E is simple.

PROPERTY 1. If E is an exceptional bundle then the following bundles are also exceptional: E^*; $E \otimes L$, where L is an element of the Picard group on $\mathbf{P}^1 \times \mathbf{P}^1$, i.e. a line bundle; $\phi^* E$, where $\phi \in \mathrm{Aut}(\mathbf{P}^1 \times \mathbf{P}^1)$.

In particular i^*E is exceptional, where $i : (x, y) \mapsto (y, x)$ is an involution of the quadric. What is more, if ϕ is an automorphism which does not switch the generating lines on the quadric

i.e. $\qquad \phi : (x, y) \mapsto (g_1 x, g_2 y), \qquad g_1, g_2 \in \mathrm{PGL}(2)$

then $\phi^* E \equiv E$, in other words E is a homogeneous bundle.

This follows from the obvious identities

$$\mathcal{E}nd(E^*) = \mathcal{E}nd(E \otimes L) = \mathcal{E}nd(\phi^* E) = \mathcal{E}nd(E)$$
$$\mathrm{Ext}^i(E, E) = \mathrm{H}^i(\mathcal{E}nd\, E).$$

E is homogeneous because it is rigid, i.e. there are no nontrivial deformations. Property 1 shows that given any exceptional bundle E it is possible to construct a whole class of exceptional bundles. Consider $E \otimes L$, $E^* \otimes L$, $i^*E \otimes L$, where L is a one-dimensional bundle and i is an involution. Therefore, to describe exceptional bundles it is enough to give one representative from this class. One can choose a canonical representative, for example, in the following way.

Let us denote

$$c_1(E) = \big(a(E), b(E)\big) \in \mathrm{Pic}(\mathbf{P}^1 \times \mathbf{P}^1) = \mathbf{Z} \oplus \mathbf{Z}$$
$$x(E) = \frac{a(E)}{r(E)}, \qquad y(E) = \frac{b(E)}{r(E)},$$

where $r(E)$ is the rank of E and $\big(x(E), y(E)\big) \in \mathbf{Q} \otimes \mathrm{Pic}(\mathbf{P}^1 \times \mathbf{P}^1)$. Then every class has exactly one representative which is represented by the point $\big(x(E), y(E)\big)$ in the plane $\mathbf{Q} \oplus \mathbf{Q}$, such that $0 \leqslant x(E) < 1$, $0 \leqslant y(E) \leqslant 1$, $x(E) \geqslant y(E)$, $x(E) + y(E) \leqslant 1$.

PROPERTY 2. With this notation for an exceptional bundle, we have

$$2a(E)b(E)\big(1 - r(E)\big) - r(E)^2 + 2r(E)c_2(E) = -1.$$

This is because the cohomological conditions on E imply that $\chi(E, E) = 1$, but

$$\chi(E, E) = \big(\mathrm{ch}\, E^* \cdot \mathrm{ch}\, E \cdot \mathrm{Td}(\mathbf{P}^1 \times \mathbf{P}^1)\big),$$

where $\mathrm{Td}(\mathbf{P}^1 \times \mathbf{P}^1) = \big(1, (1, 1), 1\big)$ and $\mathrm{ch}\, E = \Big(r(E), c_1(E), (\frac{1}{2}c_1^2 - c_2)(E)\Big)$.

COROLLARY.

(1) The second Chern class is determined by the rank and first Chern class

$$c_2(E) = \frac{2a(E)b(E) + r(E) + 1}{2r(E)}\big(r(E) - 1\big).$$

(2) The rank of an exceptional bundle must be odd.

(3) $2a(E)b(E) + 1$ is divisible by $r(E)$.

(4) $a(E)$ and $b(E)$ are each coprime to $r(E)$, i.e. $\dfrac{a(E)}{r(E)}$ and $\dfrac{b(E)}{r(E)}$ are irreducible.

PROPERTY 3. Exceptional bundles are μ-stable in the sense of Mumford-Takemoto with respect to the polarization determined by the anticanonical class $-K = (2,2)$.

COROLLARY. *There is at most one exceptional bundle with given $r(E)$ and $c_1(E)$.*

These important results are proved by A. L. Gorodentsev in [2].

Any one-dimensional bundle is a simple example of an exceptional bundle because $\mathcal{E}nd\, L = \mathcal{O}$ and $H^0(\mathcal{O}) = \mathbb{C}$, $H^1(\mathcal{O}) = H^2(\mathcal{O}) = 0$.

In the following section we shall exhibit some numerical invariants of exceptional bundles. The basic method of obtaining these invariants will be to restrict bundles to lines.

2. Constructing Short Exact Sequences from Exceptional Bundles.

THEOREM. *Let E be a simple rigid bundle of rank r on a surface X and let ℓ be a line in X with $\ell + K_X < 0$. Then*

$$E|_\ell = (r - n)\mathcal{O}(d) \oplus n\mathcal{O}(d - 1).$$

PROOF. Consider $\mathcal{E} = \mathcal{E}nd\, E = E^* \otimes E$ and the associated exact sequence

$$0 \longrightarrow \mathcal{E}(-\ell) \longrightarrow \mathcal{E} \longrightarrow \mathcal{E}|_\ell \longrightarrow 0.$$

The long exact sequence in cohomology gives

$$H^1(\mathcal{E}) \longrightarrow H^1(\mathcal{E}|_\ell) \longrightarrow H^2(\mathcal{E}(-\ell)).$$

In this sequence $H^1(\mathcal{E}) = 0$ as E is rigid. Serre duality implies

$$\left(H^2(\mathcal{E}(-\ell))\right)^* = H^0(\mathcal{E}(\ell + K_X)) = \mathrm{Hom}\big(E, E(\ell + K_X)\big).$$

But there are no nonzero homomorphisms from E to $E(\ell + K_X)$ because for $\ell + K_X < 0$ such a nonzero homomorphism would give an endomorphism of E vanishing on the σ divisor $-\ell - K_X > 0$, i.e. is not a homothety. This contradicts simplicity. Therefore $H^2(\mathcal{E}(-\ell)) = 0$ and then we have $H^1(\mathcal{E}|_\ell) = 0$, i.e. $E|_\ell$ is rigid. But any bundle on a line can be written as a sum of line bundles so $E|_\ell = \bigoplus_{i=1}^{r} \mathcal{O}(d_i)$, say. Then $\mathcal{E}|_\ell = \bigoplus_{i,j=1}^{r} \mathcal{O}(d_i - d_j)$, and $0 = H^1(\mathcal{E}|_\ell) = \bigoplus_{i,j=1}^{r} H^1(\mathcal{O}(d_i - d_j))$. Hence, $d_i - d_j \geqslant -1$ for all i, j. Let $d = \max d_i$. Then for any i either $d_i = d$ or $d_i = d - 1$. This proves the theorem.

Further on we shall prove the following

PROPOSITION. *Associated to any exceptional bundle E satisfying*

$$0 \leqslant x(E) < 1, \quad 0 \leqslant y(E) < 1, \quad x(E) + y(E) \leqslant 1$$

there is a short exact sequence of the form

$$0 \to r_4\mathcal{O}(1, -d+1) \oplus r_3\mathcal{O}(1, -d) \to E \to r_2\mathcal{O}(0, -d'+1) \oplus r_1\mathcal{O}(0, -d) \to 0,$$

where the numbers r_1, r_2, r_3, r_4, d, d' are uniquely determined by the bundle E.

In the rest of the paper we fix the following conventions $r(E) = r$, $a(E) = a$, $b(E) = b$, π_i is the projection of $\mathbf{P}^1 \times \mathbf{P}^1$ onto the corresponding factor of the direct product, which we denote by ℓ_i, for $i = 1, 2$. Then, in accordance with the theorem, for a bundle satisfying the assumptions of the proposition, we have $h^0(E|_{\ell_2}) = r - a$ and $E^*|_{\ell_2} = (r - a)\mathcal{O} \oplus a\mathcal{O}(-1)$. So $\mathbf{R}^0\pi_{1*}E^*$ is a bundle of rank $r - a$ over ℓ_1. Consider the canonical homomorphism $\pi^*(\mathbf{R}^0\pi_{1*}E^*) \xrightarrow{\text{can}} E$. Since its restriction to the first factor is the inclusion

$$\pi_1^*(\mathbf{R}^0\pi_{1*}E^*)|_{a \times \mathbf{P}^1} \longrightarrow E^*|_{a \times \mathbf{P}^1}$$

for any line $a \times \mathbf{P}^1$, there exists an exact sequence of bundles:

$$0 \longrightarrow \pi_1^*(\mathbf{R}^0\pi_{1*}E^*) \longrightarrow E^* \longrightarrow G \longrightarrow 0.$$

And also $G|_{a \times \mathbf{P}^1} = a\mathcal{O}(-1)$, i.e. $G(0,1)|_{a \times \mathbf{P}^1} = a\mathcal{O}$. Therefore $G(1,0) = \pi_1^*G'$, where $G' = \mathbf{R}^0\pi_{1*}G(0,1)$.

Now write

$$\mathbf{R}^0\pi_{1*}E^* = \bigoplus_{i=1}^{r-a} \mathcal{O}(d_i'), \qquad G' = \bigoplus_{i=1}^{q} \mathcal{O}(d_i).$$

Then $F = \pi_1^*(\mathbf{R}^0\pi_{1*}E^*) = \oplus\mathcal{O}(d_i', 0)$ and $G = \pi_1^*(G') \otimes \mathcal{O}(0, -1) = \bigoplus \mathcal{O}(d_i, -1)$. This implies that $^i\langle F \mid G \rangle = 0$, because $^i\langle F|_{\ell_2} \mid G|_{\ell_2}\rangle = 0$. In [3] Mukai proves that if $^i\langle F \mid G \rangle = 0$ then for a short exact sequence $0 \to F \to E^* \to G \to 0$ the following inequality holds

$$\dim {}^1\langle E^* \mid E^* \rangle \geqslant \dim {}^1\langle F \mid F \rangle + \dim {}^1\langle G \mid G \rangle.$$

Since $^1\langle E^* \mid E^* \rangle = 0$ we have $^1\langle F \mid F \rangle = {}^1\langle G \mid G \rangle = 0$. Therefore, exactly as in the proof of the theorem, we must have $d_i' - d_j' \geqslant -1$, $d_i - d_j \geqslant 1$ for all i, j. And so we get

$$F = r_1\mathcal{O}(d', 0) \oplus r_2\mathcal{O}(d' - 1, 0), \qquad G = r_3\mathcal{O}(d, -1) \oplus \mathcal{O}(d - 1, -1),$$

where $d' = \max d_i'$ and $d = \min d_i + 1$. Note that if $r > 1$ then such a choice of d', d ensures that we always have $r_1 \neq 0$, $r_4 \neq 0$. Substituting the explicit form of F and G into the short exact sequence and taking the dual we obtain the desired sequence.

We will now see which values the numbers d, d', r_1, r_2, r_3, r_4 can take. For the short exact sequence

$$0 \longrightarrow r_1\mathcal{O}(d',0) \oplus r_2\mathcal{O}(d'-1,0) \longrightarrow E^* \longrightarrow r_3\mathcal{O}(d,-1) \oplus r_4\mathcal{O}(d-1,-1) \longrightarrow 0$$

we have the equality $\operatorname{ch} E^* = \operatorname{ch} F + \operatorname{ch} G$. A calculation shows that

$$\operatorname{ch} E^* = \big(r, (-a, -b), ab - c_2(E)\big)$$
$$\operatorname{ch} F = \big(r_1 + r_2, (0, r_1 d' + r_2 d' - r_2), 0\big)$$
$$\operatorname{ch} G = \big(r_3 + r_4, (-r_3 - r_4, r_3 d + r_4 d - r_4), r_4 - r_4 d - r_3 d\big).$$

As in §1 we will write the condition $\chi(E, E) = 1$ as

$$c_2(E) = (r-1)\frac{r+1+2ab}{2r}. \tag{1}$$

We now have the following system of equations

$$\begin{aligned}
r_1 + r_2 &= r - a \\
r_3 + r_4 &= a \\
ad &= c_2 + r_4 - ab \\
(r-a)d' + c_2 + b - ab &= r_2 \\
c_2(E) = c_2 &= (r-1)\frac{r+1+2ab}{2r}
\end{aligned} \tag{2}$$

We will show that if $r_1 \geqslant 1$, $r_2 \geqslant 0$, $r_3 \geqslant 0$, $r_4 \geqslant 1$ then (2) has a unique solution, i.e. the triple r, a, b determines c_2, r_1, r_2, r_3, r_4, d, d'.

Rewrite the third equation in the form $r_3 = c_2 + a - ab - ad$. Since $0 \leqslant r_3 \leqslant a-1$, we see that

$$r_3 \equiv c_2 \ (\operatorname{mod} a) \qquad r_4 \equiv a - c_2 \ (\operatorname{mod} a), \tag{3}$$

and also

$$d = \frac{c_2 + r_4 - ab}{a} = \left[\frac{c_2}{a}\right] + 1 - b$$

$$\text{i.e.} \qquad d = \left[\frac{c_2 + a - ab}{a}\right], \tag{4}$$

where $[\ ]$ denotes the integer part. Similarly, from $0 \leqslant r_2 \leqslant r - a - 1$ and the fourth equation we have

$$r_2 = (c_2 + b - ab) \ (\operatorname{mod}(r-a)) \qquad r_1 = r - a - r_2 \tag{5}$$

$$d' = -\left[\frac{c_2 + b - ab}{r-a}\right]. \tag{6}$$

Equations (3), (4), (5) and (6) solve the equations in (2). This proves the proposition.

We will now see what restrictions are placed on d' and d, and hence also on r, a, b, by assuming that E is stable. Since we are considering bundles which satisfy the conditions of the proposition, $\mu(E) = \dfrac{c_1(E)}{r(E)}.(1,1) = x(E) + y(E)$ is within the limits $0 < \mu(E) \leqslant 1$. Since E has a subbundle $\mathcal{O}(1, -d+1)$ and a quotient bundle $\mathcal{O}(0, -d')$, a necessary condition is that $d' \leqslant -1$, $d \geqslant 2$.

In the subsequent calculations we will show that $d' = -1$ and if $a \geqslant b$ then $d = 2$.

We will transform (4) and (6) using the explicit form of c_2 from (1).

$$d = \left[\frac{c_2 + a - ab}{a}\right] = \left[\frac{1}{a}\left(\frac{r^2 - 1}{2r} + a - \frac{ab}{r}\right)\right] = \left[\frac{r^2 - 1}{2ar} - \frac{b}{r}\right] + 1$$

$$-d' = \left[\frac{c_2 + b - ab}{r - a}\right] = \left[\frac{1}{r - a}\left(\frac{r^2 - 1}{2r} + b - \frac{ab}{r}\right)\right] = \left[\frac{r^2 - 1}{2r(r - a)} + \frac{b}{r}\right].$$

From which we get

$$\frac{r^2 - 1}{2ar} - \frac{b}{r} \geqslant d - 1 \qquad \frac{r^2 - 1}{2r(r - a)} + \frac{b}{r} \geqslant -d', \quad \text{respectively.}$$

Rearranging these we have

$$r^2 - 1 \geqslant ((d - 1)r + b).2a, \qquad r^2 - 1 \geqslant (-d'r - b).2(r - a). \tag{8}$$

We already know that $d - 1 \geqslant 1$, $-d' \geqslant 1$, i.e. we get

$$r^2 - 1 \geqslant (r + b)2a \qquad r^2 - 1 \geqslant (r - b)2(r - a). \tag{9}$$

Putting these inequalities together we get

$$r^2 - 1 \geqslant a(r + b) + (r - a)(r - b) = r^2 + 2ab - br$$

or $(2a - r)b \leqslant -1 < 0$ which gives $a < \frac{r}{2}$. If we take the canonical short exact sequence for the bundle i^*E, where i is the involution on the quadric then the same argument also shows that $b < \frac{r}{2}$. If we now assume that $d' \leqslant -2$, then the second inequality of (9) implies that

$$r^2 - 1 \geqslant (2r - b)2(r - a) > \frac{3r^2}{2},$$

which is impossible. This means that for all bundles $d' = -1$.

We now assume that $a \geqslant b$. Then we see from the second inequality of (9) that $r^2 > r^2 - 1 \geqslant 2(r - a)^2$, i.e. $r > \sqrt{2}(r - a)$, which implies that $a > (1 - \sqrt{2}/2)r$. Then the first inequality of (8) implies that

$$r^2 > ((d - 1)r + b)2a > (d - 1)r(2 - \sqrt{2})r,$$

which shows that $d - 1 < \dfrac{1}{2 - \sqrt{2}} = \dfrac{2 + \sqrt{2}}{2} < 2$, i.e. $d = 2$.

It is easy to solve system (2) with the fixed values $d' = -1$ and $d = 2$ for $a \geqslant b$. Then we get

$$r_1 = r + \frac{ab}{r} + \frac{r^2 + 1}{2r} - 2a - b \qquad r_3 = \frac{r^2 - 1}{2r} - \frac{ab}{r} - a$$
$$r_2 = r - a - r_1 \qquad r_4 = a - r_3.$$

REMARK. If we do not assume $a \geqslant b$ then there exist bundles for which d can take any value bigger than two.

It would be interesting to consider the question of which possible values r, a, b, r_1, r_2, r_3, r_4, c_2 can take when E is exceptional and which cocycles give an exceptional bundle in the canonical short exact sequence. In connection with this problem we will describe the homological interpretation of the numbers r_1, r_2, r_3, r_4. If we apply the following functors to the canonical short exact sequence

$$|\mathcal{O}(-2,-3)\rangle, \quad |\mathcal{O}(-2,-2)\rangle, \quad |\mathcal{O}(-1,1)\rangle, \quad |\mathcal{O}(-1,0)\rangle.$$

Then we find

$$r_1 = {}^2\langle E \mid \mathcal{O}(-2,-3)\rangle, \qquad r_2 = {}^1\langle E \mid \mathcal{O}(-2,-2)\rangle,$$
$$r_3 = {}^0\langle E \mid \mathcal{O}(-1,1)\rangle, \qquad r_4 = {}^1\langle E \mid \mathcal{O}(-1,0)\rangle.$$

3. Restricting Exceptional Bundles to Elliptic Curves.

LEMMA. *Let C be a smooth elliptic curve in $\mathbf{P}^1 \times \mathbf{P}^1$ and E an exceptional bundle on $\mathbf{P}^1 \times \mathbf{P}^1$. Then*

$$H^0\big(\mathrm{Hom}(E|_C, E|_C)\big) = H^1\big(\mathrm{Hom}(E|_C, E|_C)\big) = \mathbf{C}.$$

PROOF. We tensor the structure sequence of C with the sheaf $\mathcal{E}nd\, E$ to obtain

$$0 \longrightarrow (\mathcal{E}nd\, E) \otimes K \longrightarrow \mathcal{E}nd\, E \longrightarrow \mathcal{E}nd\, E|_C \longrightarrow 0.$$

Then the corresponding long exact sequence in cohomology is

$$\mathbf{C}$$
$$\|$$

$$0 \longrightarrow H^0(\mathcal{E}nd\, E) \longrightarrow H^0(\mathcal{E}nd\, E|_C) \longrightarrow H^1\big((\mathcal{E}nd\, E) \otimes K\big) \longrightarrow$$

$$\longrightarrow 0 \longrightarrow H^1(\mathcal{E}nd\, E|_C) \longrightarrow H^2\big((\mathcal{E}nd) \otimes K\big) \longrightarrow 0.$$

But Serre duality implies

$$H^1\big((\mathcal{E}nd\,E)\otimes K\big)^* = H^1(\mathcal{E}nd\,E) = 0,$$
$$H^2\big((\mathcal{E}nd\,E)\otimes K\big)^* = H^0(\mathcal{E}nd\,E) = \mathbf{C},$$

which proves the lemma.

COROLLARY. *Let E be an exceptional bundle. Then $a(E)+b(E)$ and r are coprime.*

PROOF. Atiyah in [1] proves that a bundle on an elliptic curve is simple only if $\deg E$ and $r(E)$ are coprime. But if E is an exceptional bundle, $\deg E|_C = 2\big(a(E)+b(E)\big) = C.c_1(E)$. Then the lemma and the fact that the rank of E is odd gives us what we need.

COROLLARY. *Let E and F be exceptional bundles such that $0 \leqslant x(E) < 1$, $0 \leqslant y(E) < 1$, $0 \leqslant x(F) < 1$, $0 \leqslant y(F) < 1$, $x(E) \leqslant y(E)$. Then the following are equivalent:*

(1) $\mathrm{Ext}^l(E,F) = \mathrm{Ext}^l(F,E) = 0$ *for all l*

(2) $c_1(E) = (a, a+1)$ *and* $E = i^*F$,   *where i is the involution on the quadric.*

PROOF. From Riemann-Roch

$$\chi(E,F) = r(E)r(F)\left[\left(\frac{a(F)}{r(F)} - \frac{a(E)}{r(E)}\right)\left(\frac{b(F)}{r(F)} - \frac{b(E)}{r(E)}\right)\right.$$
$$\left. -\mu(F) + \mu(E) + \delta(F) + \delta(E)\right],$$

where $\mu(F) = \dfrac{c_1(F)}{r(F)}.(1,1) = \dfrac{a(F)+b(F)}{r(F)}$ and for exceptional bundles

$$\delta(F) = \frac{1}{2r^2(E)}$$

Suppose (1) holds. Then

$$0 = \chi(E,F) + \chi(F,E) = 2\big(\mu(E) - \mu(F)\big)r(E)r(F).$$

Therefore $\mu(E) = \mu(F)$ and from the first corollary $\mu(E) = \dfrac{a(E)+b(E)}{r(E)}$ is an irreducible fraction, i.e. $r(E) = r(F)$ and $a(E) + b(E) = a(F) + b(F)$. Let $c_1(E) = (a, b)$, $c_1(F) = (a - x, b + x)$. Then $\frac{x}{r}\left(-\frac{x}{r}\right) + \frac{1}{r^2} = \chi(E,F) = 0$ i.e. $x = \pm 1$. Because of the symmetry of the problem we can put $x = 1$. As we noted before

$$2ab \equiv 1 \pmod{r} \qquad 2(a+1)(b-1) \equiv 1 \pmod{r}$$

hence $b - a \equiv 1 \pmod{r}$, but by assumption $r > b \geqslant a \geqslant 0$, consequently $b = a + 1$ and so $E = i^*F$.

Conversely suppose that (2) holds. Then

$$\chi(E, i^*E) = r^2 \left(-\frac{1}{r^2} + \frac{1}{r^2} \right) = 0.$$

But $\mu(E) = \mu(i^*E)$ and stability implies that $\operatorname{Hom}(E, i^*E) = 0$ and then

$$\operatorname{Ext}^2(E, i^*E) = \operatorname{Hom}(i^*E, E \otimes K) = 0.$$

The above equalities immediately imply that $\operatorname{Ext}^1(E, i^*E) = 0$. By symmetry, we can also deduce that $\operatorname{Ext}^l(i^*E, E) = 0$. This completes the proof.

References

[1] ATIYAH, M.F., Vector Bundles Over an Elliptic Curve, *Proc. Lond. Math. Soc,* **VII** (1957) 414–452.

[2] GORODENTSEV, A.L., Exceptional Bundles on Surfaces with a Moving Anti-canonical Class, *Math. USSR Isv.,* **33** (1989) 67–83.

[3] MUKAI, S., On the Moduli Spaces of Bundles on K3 Surfaces, I, in *Vector Bundles* ed. Atiyah et al, Oxford Univ. Press, Bombay, (1986) 341–413.

4. Exceptional Bundles of Small Rank on $\mathbf{P}^1 \times \mathbf{P}^1$

D.Yu. Nogin

All exceptional bundles on $\mathbf{P}^2$ can be obtained as mutations of a helix [2]. Since an exceptional bundle E is stable, it is determined by its rank $r(E)$ and its determinant $c_1(E)$. Hence the set of exceptional bundles on $\mathbf{P}^2$ can be described by giving all possible pairs (r, c_1). This was done in [3]. Rudakov has proved in [4] that all exceptional bundles on $\mathbf{P}^1 \times \mathbf{P}^1$ can be obtained by mutations of a helix with foundation

$$\mathcal{O}, \; \mathcal{O}(1,0), \; \mathcal{O}(0,1), \; \mathcal{O}(1,1). \tag{1}$$

Exceptional bundles on $\mathbf{P}^1 \times \mathbf{P}^1$ are also stable and are again described by the pair (r, c_1). It is not known, however, what the set of all such pairs is. In this paper we shall list all such pairs for $r < 200$.

Let $a(E)$ and $b(E)$ denote components of the first Chern class

$$c_1(E) = \big(a(E), b(E)\big).$$

By using the operations of restriction to line bundles and forming the dual, we can reduce every exceptional bundle to a bundle E satisfying

$$0 \leqslant a(E) < r(E), \quad 0 \leqslant b(E) < r(E), \quad a(E) + b(E) \leqslant r(E). \tag{2}$$

There is an involution of $\mathbf{P}^1 \times \mathbf{P}^1$ which sends the bundle E to the bundle $\tilde{E}$ for which $c_1(\tilde{E}) = \big(b(E), a(E)\big)$. Therefore it suffices to describe all the exceptional bundles for which

$$a(E) \geqslant b(E). \tag{3}$$

We introduce the notation

$$\lambda(E) = \frac{c_1(E)}{r(E)} \in \mathrm{Pic}(\mathbf{P}^1 \times \mathbf{P}^1) \otimes \mathbf{Q}$$

$$\rho(E) = \frac{c_1^2(E) - 2c_2(E)}{2r(E)} \in \mathbf{Q}.$$

Then the Chern character satisfies $\mathrm{ch}(E) = r(E).\big(1, \lambda(E), \rho(E)\big)$ and the Riemann-Roch Theorem implies

$$\chi(E, F) = r(E).r(F)\Big(1 - \frac{1}{2}(\lambda(F) - \lambda(E)).K + \rho(E) + \rho(F) - \lambda(E)\lambda(F)\Big), \tag{4}$$

where K is the canonical class of $\mathbf{P}^1 \times \mathbf{P}^1$, equal to $\mathcal{O}(-2, -2)$. The quantity

$$\mu(E) = -\frac{1}{2}\lambda(E).K$$

is the *slope* of the bundle E. In the case of an exceptional bundle E, (4) implies

$$r^2(E)\left(1 + 2\rho(E) - \lambda^2(E)\right) = \chi(E, E) = 1.$$

If we determine $\rho(E)$ from this formula and substitute the result in (4) we obtain the following Riemann-Roch formula for exceptional bundles

$$\chi(E, F) = r(E).r(F)\left[\left(\mu(F) - \mu(E)\right) + \left(x(E) - x(F)\right)\left(y(E) - y(F)\right) + \right.$$
$$\left. + \frac{1}{2r^2(E)} + \frac{1}{2r^2(F)}\right], \tag{5}$$

where $x(E) = \dfrac{a(E)}{r(E)}$, $y(E) = \dfrac{b(E)}{r(E)}$, i.e. $\lambda(E) = \left(x(E), y(E)\right)$ and therefore

$$\frac{1}{2}\left(\lambda(E) - \lambda(F)\right)^2 = \left(x(E) - x(F)\right)\left(y(E) - y(F)\right).$$

As was shown in [1] a necessary condition for E to be exceptional is

$$r(E) \mid \left(2a(E)b(E) + 1\right). \tag{6}$$

It is evident from this condition that the rank of an exceptional bundle on $\mathbf{P}^1 \times \mathbf{P}^1$ is odd and the fractions $a(E)/r(E)$ and $b(E)/r(E)$ are irreducible, in other words the point $\left(x(E), y(E)\right)$ uniquely determines $r(E)$ and $c_1(E)$, and hence, the exceptional bundle. Therefore one can identify an exceptional bundle with the point representing it. This approach turned out to be very useful in [4].

The general properties of stable bundles with the condition $\mu(F) < \mu(E)$ imply that there are no non-trivial homomorphisms from E to F. In the case of exceptional bundles on $\mathbf{P}^1 \times \mathbf{P}^1$ with $\mu(F) = \mu(E)$ all the non-trivial homomorphisms are isomorphisms. In other words, if $\mu(F) \leqslant \mu(E)$ and $F \neq E$ then ${}^0\langle E \mid F \rangle = 0$. If we want $\mu(E(K)) < \mu(F)$, then we must have ${}^0\langle F \mid E(K) \rangle = 0$. But Serre duality gives

$$\left({}^0\langle F \mid E(K) \rangle\right)^* = {}^2\langle E \mid F \rangle.$$

This means that if $\mu(E(k)) = \mu(E) - 4 < \mu(F) \leqslant \mu(E)$ then we must have

$$\chi(E, F) = -{}^1\langle E \mid F \rangle \leqslant 0.$$

Let us fix a bundle E. The remark above imposes the same restrictions on $(x(F), y(F)) = (x, y)$. When $\mu(E) - 4 < x + y < \mu(E)$ we see from (5) that

$$\mu(F) - \mu(E) + \big(x - x(E)\big)\big(y - y(E)\big) + \frac{1}{2r^2(E)} + \frac{1}{2r^2(F)} \leq 0,$$

or

$$x + y - x(E) - y(E) + \big(x - x(E)\big)\big(y - y(E)\big) + 1 \leq 1 - \frac{1}{2r^2(E)} - \frac{1}{2r^2(F)}.$$

By throwing away the term $\dfrac{1}{2r^2(E)}$, we obtain a necessary condition on (x, y)

$$\big((x - x(E) + 1)\big(y - y(E) + 1\big) < 1 - \frac{1}{2r^2(E)}, \qquad \text{where} \quad \mu(E) - 4 < x + y \leq \mu(E). \quad (7)$$

This condition has to be satisfied by all the x, y representing any exceptional bundle which is not isomorphic to E.

Inequality (7) defines the set of points of the plane (x, y) which are bounded by a hyperbola and two straight lines. By an analogous inequality for $E(2, 2)$, we see that every exceptional bundle E is enclosed in a crescent shaped region inside which there are no exceptional bundles other than E. This crescent is bounded by the branches of the two hyperbolae and is determined by the following equalities

$$\begin{cases} \big(x - x(E) + 1\big)\big(y - y(E) + 1\big) \geq 1 - \dfrac{1}{2r^2(E)} & \text{where } \mu(E) - 4 < x + y \leq \mu(E) \\[2mm] \big(x - x(E) - 1\big)\big(y - y(E) - 1\big) \geq 1 - \dfrac{1}{2r^2(E)} & \text{where } \mu(E) - 4 \geq x + y > \mu(E). \end{cases}$$
$$(8)$$

Note that (x, y) satisfy (8) and so E lies in its corresponding crescent and hence every crescent is non-empty. As each exceptional bundle lies in its own crescent, conditions (8) are independent for different E.

By mutating helix (1), one can obtain various collections of exceptional bundles. In order to prove that we get all the exceptional bundles of a given rank r (up to restriction, taking the dual and switching the generators of the base space), it is enough to consider the rational points with parameter r lying in the triangle defined by inequalities (2) and (3), and to ignore those which do not satisfy (6) or lie in the crescents of the bundles already constructed. This has been done for all $r < 200$ and hence we have a description of all the exceptional bundles of such ranks. The results are contained in the table at the end of this paper.

Let $\begin{Bmatrix} \alpha_F E \\ \gamma_F E \\ \varepsilon_F E \end{Bmatrix}$ denote the result of a $\begin{Bmatrix} \text{division} \\ \text{recoil} \\ \text{extension} \end{Bmatrix}$ type shift of the bundle E by F. Let $\tilde{E}$ denote the bundle obtained from E by switching the generators of the base space.

In the tables we mention the numbers r_1, r_2 defined by the short exact sequence

$$0 \longrightarrow r_4 \mathcal{O}(1, -d+1) \oplus r_3 \mathcal{O}(1, -d) \longrightarrow E \longrightarrow r_2 \mathcal{O}(0, 2) \oplus r_1 \mathcal{O}(0, 1) \longrightarrow 0. \qquad (9)$$

The existence of this is proved in the previous paper in this collection. The other numbers are given by $r_1 = r - a - r_2$, $r_4 = a - r_3$, and $d = 2$ when $a \geqslant b$. The numbers $\tilde{r}_2$, $\tilde{r}_3$, d are determined by sequence (9) for $\tilde{E}$. Here $\tilde{r}_1 = r - b - \tilde{r}_2$, $r_4 = b - \tilde{r}_3$.

In table 1 general characteristics of some series of the bundles E_n, A_n^+, A_n^-, B_n, D_n are given. In table 2 all bundles other than E_n, A_n^+ and A_n^- are given.

References

[1] GORODENTSEV, A.L., Transformations of Exceptional Bundles on $\mathbf{P}^n$, *Math. USSR Isv.*, **32** (1989) 1–13.

[2] GORODENTSEV, A.L., Exceptional Bundles on Surfaces with a Moving Anticanonical Class, *Math. USSR Isv.*, **33** (1989) 67–83.

[3] RUDAKOV, A.N., The Markov Numbers and Exceptional Bundles on $\mathbf{P}^2$, *Math. USSR Isv.*, **32** (1989) 99–112.

[4] RUDAKOV, A.N., Exceptional Bundles on a Quadric, *Math. USSR Isv.*, **33** (1989) 115–138.

Table 1.

r	a,b	r_2, r_3	$\tilde{d};\ \tilde{r}_2, \tilde{r}_3$	construction
$2n+1$, $n \geqslant 2$	$n, 1$	$0, 0$	$n+1;\ 0, 0$	$E_n = \varepsilon_{\mathcal{O}(1,0)}\mathcal{O}(-n,1)$ $= \alpha_{\mathcal{O}}\mathcal{O}(-n,-1)$ $= \alpha_{E_{n-1}}E_{n-2}$
$12n+5$, $n \geqslant 0$	$4n+1,\ 3n+2$	$0, n+1$ for $n \geqslant 1$	$2;\ 0, 2n$	$A_n^- = \alpha_{A_{n-1}^-}A_{n-2}^-$, $A_0^- = \widetilde{E}_2,\ A_1^- = \varepsilon_{\widetilde{E}_2}E_3$
$12n+7$, $n \geqslant 0$	$4n+3,\ 3n+1$	$0, n$	$2;\ 0, 2n+2$ for $n \geqslant 2$	$A_n^+ = \alpha_{A_{n-1}^+}A_{n-2}^+$, $A_0^+ = E_3,\ A_1^+ = \alpha_{E_2}\mathcal{O}(1,0)$
$4n^2+2n-1$, $n \geqslant 1$	$2n^2-1,\ 2n$	$0, 1$ for $n \geqslant 2$	$n+1;\ 0, 0$	$B_n = \alpha_{E_n}\mathcal{O}(1,0)$
$4n^2+6n+1$, $n \geqslant 1$	$2n^2+2n,\ 2n+2$	$1, 0$	$n+1;\ 1, 0$	$D_n = \alpha_{E_n}\mathcal{O}$

Table 2.

r	a,b	r_2, r_3	$\tilde{d}; \tilde{r}_2, \tilde{r}_3$	construction
11	4,4	1,0	2; 1,0	D_1
19	7,4	0,1	3; 0,0	B_2
29	12,6	1,0	3; 1,0	D_2
41	17,6	0,1	4; 0,0	B_3
	15,15	4,0	2; 4,0	$\alpha_{D_1} E_1$
55	24,8	1,0	4; 1,0	D_3
59	18,18	1,6	2; 1,6	$\alpha_{E_1} \mathcal{O}(2,2)$
65	24,23	6,0	2; 6,1	$\alpha_{E_1} \mathcal{O}(-2,-1)$
				$= \alpha_{D_1} \mathcal{O}(0,1)$
71	31,8	0,1	5; 0,0	B_4
	26,15	0,4	3; 0,0	$\alpha_{B_2} E_2$
89	40,10	1,0	5; 1,0	D_4
99	29,29	0,12	2; 0,12	$\varepsilon_{\tilde{A}_2^-} A_3^-$
109	49,40	0,1	6; 0,0	B_5
111	34,31	0,12	2; 0,15	$\alpha_{A_2^-} E_2$
131	60,12	1,0	6; 1,0	D_5
153	56,53	15,0	2; 15,0	$\alpha_{\alpha_{D_1} E_1} D_1$
155	71,12	0,1	7; 0,0	B_6
	64,23	0,4	3; 0,22	$\alpha_{B_3} E_4$
169	70,35	6,0	3; 6,0	$\alpha_{D_2} E_2$
	66,32	1,6	3; 1,6	$\alpha_{E_2} \mathcal{O}(2,2)$
	50,49	0,20	2; 0,21	$\alpha_{\varepsilon_{\tilde{A}_2^-} A_3^-} \tilde{A}_2^-$
179	73,38	6,1	2; 6,36	$\alpha_{E_2} \mathcal{O}(-2,-1)$
181	84,14	1,0	7; 1,0	D_6
	64,41	0,12	2; 0,35	$\alpha_{A_2^+} E_2$
	55,51	0,20	2; 0,24	$\alpha_{\alpha_{A_2^-} E_2} A_3^-$

5. On the Functors Ext• Applied to Exceptional Bundles on $\mathbf{P}^2$

A. I. Bondal   A. L. Gorodentsev

In the present paper we apply helix theory, developed in [1], to describe those values of i for which the spaces $\mathrm{Ext}^i(A, B)$ do not vanish, in terms of the slopes of two arbitrary bundles A and B on $\mathbf{P}^2$. One expects such a description to exist on the basis of Proposition 5.11 in [1] (given without proof). Here we prove a stronger result than that proposition:

THEOREM. *Let A and B be exceptional bundles on $\mathbf{P}^2$. Then*

(1) if $\mu(A) \leqslant \mu(B)$, then $\mathrm{Ext}^1(A, B) = \mathrm{Ext}^2(A, B) = 0$ and the canonical map

$$\mathrm{Hom}(A, B) \otimes A \to B$$

is an epimorphism, and the canonical map

$$A \to \mathrm{Hom}(A, B)^* \otimes B$$

is a monomorphism;

(2) if $\mu(A) - 3 < \mu(B) < \mu(A)$, then $\mathrm{Hom}(A, B) = \mathrm{Ext}^2(A, B) = 0$;

(3) if $\mu(B) \leqslant \mu(A) - 3$, then $\mathrm{Hom}(A, B) = \mathrm{Ext}^1(A, B) = 0$, and $\mathrm{Ext}^2(A, B) \neq 0$.

The proof of this theorem, which will take all of section 2, is based on the graphic representation of a helix on $\mathbf{P}^2$ by the points of a Markov tree and the process of "induction on the Markov tree". Section 1 is devoted to giving a precise meaning to all these notions.

All the notation we use and the basic notions of the theory of helices on $\mathbf{P}^2$ are taken from [1] (in particular $\mu(E)$ denotes the slope of a bundle E, i.e. is the rational number $\dfrac{c_1(E)}{r(E)}$).

1.   A Graphic Representation of Helices.

1.1.   The Markov Tree.

By a *Markov tree* we mean the following Brauer-Tits graph with distinguished vertex 0:

(1)

It is an infinite tree with three edges attached to each vertex. To each vertex of the Markov tree we associate a *weight*, this is the length of the shortest path joining this vertex to the point 0. Any path on the Markov tree in which the weights of the vertices we pass through increase will be called an *outward path* and a path in the reverse direction will be called an *inward path*.

The group S_3 acts on the Markov tree (1) as the group of automorphisms preserving preserving the point 0. This action is generated by rotations of $\pm 120°$ about the point 0 and by reflections in the lines extending the edges attached to 0.

1.2.   Graphs from Helices.

It is easy to see from the results of [1] (pp 5.2–5.8) that there is a one-one correspondence between helices on $\mathbf{P}^2$ and vertices of a Markov tree, such that the point 0 corresponds to the helix of invertible sheaves $\sigma_0 = \{\mathcal{O}(i)\}_{i \in \mathbf{Z}}$, and the edges correspond to mutations used to obtain the helices, which correspond to neighbouring vertices.

A mutation of a helix is defined by a choice of the bundle (up to twisting by multiples of the canonical class) which we want to shift, and by a choice of which direction the shift will be performed—to the right or to the left. The period of a helix on $\mathbf{P}^2$ is equal to three and it follows from the definition of a helix that for any $E \in \sigma$ we have the relation $L_E \sigma = R_E \sigma$ (see the explaination of 1.10 in [1]). Therefore any helix on $\mathbf{P}^2$ has *exactly three mutations*, which correspond to edges of the Markov tree. We obtain the one-one correspondence mentioned above by successively forming new helices from σ_0 and assigning them to the vertices of the Markov tree. Here it is convenient to introduce the unique numbering of the bundles in all the helices, assuming that under a mutation the number of the bundle being shifted does not change and the sheaves $\mathcal{O}(i)$ in the helix σ_0 are numbered by i itself.

By identifying the action of the group S_3 on the Markov tree we get an action of this group on the set of helices. Then the rotation by $120°$ around the point 0 corresponds to twisting by $\mathcal{O}(1)$ and the reflection in the line containing the edge given by shifting the bundle $\mathcal{O}$ in the helix σ_0, corresponds to dualizing.

1.3. Convenient Foundations.

The foundation (A, B, C) of a helix on $\mathbf{P}^2$ will be called *convenient* if

$$\mu(C) - \mu(A) \leqslant 1.$$

It is clear that amongst the three different foundations (modulo twists) $(B, C, A(3))$, $(C, A(3), B(3))$ of an arbitrary helix σ only one can be convenient. It is also clear that the unique helix which has no convenient foundations is the initial helix σ_0. All the helices obtained from σ_0 on some outward motion on the Markov tree, the first mutation of which consists of a shift of $\mathcal{O}$, $\mathcal{O}(1)$ or $\mathcal{O}(2)$ have convenient foundations which lie in intervals $[1 + 3n, 2 + 3n]$, $[2 + 3n, 3 + 3n]$ and $[0 + 3n, 1 + 3n]$, $n \in \mathbf{Z}$, respectively. Moreover, if (A, B, C) is a convenient foundation of a helix σ, then an outward motion from σ leads to the helices with convenient foundations $(B, R_\sigma A, C)$ and $(A, L_\sigma C, B)$. This motivates the following definition.

1.4. Definition.

Let (A, B, C) be any foundation of a helix σ on $\mathbf{P}^2$. By *internal mutations* of this foundation we mean foundations $(B, R_\sigma A, C)$ and $(A, L_\sigma C, B)$, i.e. foundations which are obtained by a shift of one of the boundary bundles inwards (then the middle bundle goes to the boundary). On the other hand an *external mutation* means a shift of the middle bundle to the boundary. In this way, the remark in the previous section means that moving outwards on the Markov tree corresponds to an internal mutation of convenient foundations and moving inwards on the Markov tree corresponds to an external mutation of convenient foundations.

1.5. The Classification of Exceptional Bundles on $\mathbf{P}^2$.

According to §5.10 in paper [1], every exceptional bundle E on $\mathbf{P}^2$ is contained in some helix which is constructed in the following way (see also the paper of Rudakov [2]). Let $k < \mu(E) < k + 1$. We will consider the convenient foundation

$$\left(\mathcal{O}(k), \mathcal{T}(k-1), \mathcal{O}(k+1) \right)$$

of the helix $L_{\mathcal{O}(k+2)}\sigma_0 = R_{\mathcal{O}(k-1)}\sigma_0$ and we will divide up its mutations so that $\mu(E)$ stays squeezed between the slopes of the boundary bundles of the newly obtained foundation, i.e. we will move outwards on the Markov tree moving from the convenient foundation (A, B, C) to the foundation (B, RA, C) if $\mu(B) < \mu(E) < \mu(C)$, and to the foundation (A, LC, B) if $\mu(A) < \mu(E) < \mu(B)$. Then after a finite number of steps we come to the convenient foundation (A', B', C') with $B' = E$. The basic principle of "induction on the Markov tree" is the following proposition which makes it possible to dispense with the fact that the foundations which appear in the above algorithm are convenient.

1.6. Proposition.

Let (F_1, F_2, F_3) be any foundation of an arbitrary helix on $\mathbf{P}^2$, and let E be an exceptional bundle for which $\mu(F_1) \leqslant \mu(E) \leqslant \mu(F_3)$. Then via internal mutations of the foundation (F_1, F_2, F_3) one can construct a foundation (F_1', F_2', F_3') with $F_2' = E$.

PROOF. If the initial foundation is convenient then there is nothing to prove. If not, then we will apply the algorithm given in 1.5, but remembering that the intermediate foundations (A, B, C) may not be convenient. Suppose $\mu(A) < \mu(E) < \mu(B)$ and we pass to the foundation (A, LC, B). Since one of the foundations (A, B, C), $(C(-3), A, B)$, $(B, C, A(3))$ is convenient, either the new foundation we get is convenient (and there is nothing to prove) or our mutation corresponds to an inward motion on the Markov tree. Analysing in a similar way the case $\mu(B) < \mu(E) < \mu(C)$, we see that in the case when (A, B, C) is not convenient, one step in our algorithm either takes us to a convenient foundation or moves us inwards on the Markov tree. In this way, after a finite number of steps, we will reach the effective range of the algorithm in 1.5, which proves the proposition.

2. Proof of the Theorem.

First of all we note that statement (3) follows from (1) by Serre duality and statement (2) is a simple consequence of the *stability* of exceptional sheaves on $\mathbf{P}^2$. From stability (see [1] 4.1) and from our inequalities on the slope, it follows that $\mathrm{Hom}(A, B) = 0$ and $\mathrm{Hom}(B, A(-3)) = 0$, and from Serre duality that $\mathrm{Ext}^2(A, B) = 0$.

Thus it is enough to prove statement (1) and moreover, since the morphism $A \to \mathrm{Hom}(A, B)^* \otimes B$ from the pair (A, B) is dual to the morphism $\mathrm{Hom}(B^*, A^*) \otimes B^* \to A^*$ from the pair (B^*, A^*), it suffices to prove the epimorphism condition only—the monomorphism condition follows from this.

We need two lemmas.

2.1. Lemma.

Let $\varepsilon = (A, B)$ be an exceptional pair and let E be an exceptional bundle such that

$$\mathrm{Ext}^1(E, B) = \mathrm{Ext}^2(E, B) = \mathrm{Ext}^2(E, A) = 0.$$

Then $\mathrm{Ext}^1(E, R_\varepsilon A) = \mathrm{Ext}^2(E, R_\varepsilon A) = 0.$

PROOF. By definition, the bundle $R_\varepsilon A$ is given by the short exact sequence

$$0 \longrightarrow A \longrightarrow V \otimes B \longrightarrow R_\varepsilon A \longrightarrow 0,$$

where $V = \mathrm{Hom}(A, B)^*$. Applying $\mathrm{Hom}(E, *)$ to it, we get the long exact sequence

$$0 \to \mathrm{Hom}(E, A) \to V \otimes \mathrm{Hom}(E, B) \to \mathrm{Hom}(E, R_\varepsilon A) \to \mathrm{Ext}^1(E, A) \to 0$$

which continues to the right with zeros.

2.2. Lemma.

Let $\varepsilon = (A, B)$ be an exceptional pair and let E be any bundle for which the canonical morphism $\mathrm{Hom}(E, B) \otimes E \to B$ is an epimorphism. Then the morphism $\mathrm{Hom}(E, R_\varepsilon A) \otimes E \to R_\varepsilon A$ is also an epimorphism.

PROOF. This is immediate if we consider the following commutative diagram:

$$
\begin{array}{ccc}
0 & & \\
\uparrow & & \\
\mathrm{Hom}(A, B)^* \otimes B & \xrightarrow{\ \mathrm{can}\ } & R_\varepsilon A \longrightarrow 0 \\
{\scriptstyle \mathrm{Id}\otimes\mathrm{can}}\uparrow & & \uparrow{\scriptstyle \mathrm{can}} \\
\mathrm{Hom}(A, B)^* \otimes \mathrm{Hom}(E, B) \otimes E & \xrightarrow{\ \mathrm{can}\otimes\mathrm{Id}\ } & \mathrm{Hom}(E, R_\varepsilon A) \otimes E.
\end{array}
$$

2.3.

Now we will deduce statement (1) of the proposition assuming that A and B lie in the same helix σ. Let $\sigma = \{E_i\}_{i \in \mathbf{Z}}$ and $A = E_0$. Statement (1) for $B = E_0, E_1, E_2$ follows directly from the definition of a helix, and for $B = E_i$ for arbitrary $i > 0$ it is obtained by induction on i using Lemma 2.1 and Lemma 2.2 and the fact that $E_i = R_\sigma^{(2)} E_{i-3}$.

2.4.

Finally, consider the *general case*. Let A lie in a helix σ. Choose a foundation (F_1, F_2, F_3) of the helix σ such that the inequality $\mu(F_1) \leqslant \mu(B) \leqslant \mu(F_3)$ is satisfied and F_1 does not lie further to the left than A in σ (although $A = F_1$ is entirely possible). Proposition 1.6 implies that the bundle B can be obtained by using internal mutations of the collection (F_1, F_2, F_3). For $B = F_1, F_2, F_3$ our condition holds because of 2.3. It remains to prove that if (1) holds for $B \in (Q, S, T)$, then it will hold for $B = RQ$ and for $B = LT$. To do this in the first case one must apply 2.1 and 2.2 to the pair (Q, S) and in the second case to the pair $(T(-3), Q)$.

This proves the theorem.

References

[1] GORODENTSEV, A.L. & RUDAKOV, A.N., Exceptional Vector Bundles on Projective Space, *Duke Math. J.*, **54** (1987) 115–130.

[2] RUDAKOV, A.N., The Markov Numbers and Exceptional Bundles on $\mathbf{P}^2$, *Math. USSR Isv.*, **32** (1989) 99–112.

6. Homogeneous Bundles

A.I. Bondal   M.M. Kapranov

In [4], a relationship was established between the ranks of exceptional bundles on $\mathbf{P}^2$ and solutions of the Markov equation $x^2 + y^2 + z^2 = 3xyz$. For this reason, we wish to have as explicit as possible a description of these bundles. This could be of use in connection with the Markov problem [3], which, in algebro-geometric terms, says that there exists at most one exceptional bundle on $\mathbf{P}^2$ of a given rank, up to twisting by $\mathcal{O}(i)$ and conjugation.

Since exceptional bundles on homogeneous spaces are homogeneous, it is useful to work first with arbitrary homogeneous bundles and afterwards to pick out the exceptional ones.

We will consider manifolds of the form $X = G/P$, where G is a compact semisimple algebraic group and $P \subset G$ is a parabolic subgroup. Furthermore, we will mainly be interested in the case when X is a projective space. Following [6], a *G-bundle* on X will be a bundle E together with isomorphisms $\varphi_g : g^*E \overset{\sim}{\longrightarrow} E$, defined for arbitrary $g \in G$, which are subject to the natural compatibility condition

$$\varphi_{g_1 \cdot g_2} = \varphi_{g_2} \circ g_2^* \varphi_{g_1} : g_2^* g_1^* E \to g_2^* E \to E$$

We will denote G-bundles by pairs (E, φ).

A bundle E is called *homogeneous* if it admits at least one G-bundle structure.

PROPOSITION 1. *Let (E, φ) and (F, ψ) be two G-bundles on X. Suppose that the bundles E and F are indecomposable and isomorphic (as bundles without a G-structure). Then there exists a character $\chi : G \to \mathbf{C}^*$ and an isomorphism of G-bundles $(E, \varphi \cdot \chi) \overset{\sim}{\longrightarrow} (F, \psi)$.*

PROOF. Consider the vector space $\mathrm{Hom}(E, F)$ with the natural action of G. It is a right module for the algebra $\mathrm{End}(E)$. As E is indecomposable, the semi-simple reduction of this algebra (i.e. the quotient by the radical $R \subset \mathrm{End}(E)$) is isomorphic to $\mathbf{C}$. Consequently, $L = \mathrm{Hom}(E, F) \cdot R$ is a G-invariant subspace of $\mathrm{Hom}(E, F)$ of codimension 1. Since G is semisimple, the subspace L has a canonical G-invariant complement $H \subset \mathrm{Hom}(E, F)$ with $\dim H = 1$. The action of G on H gives the character we are looking for.

In this way, the study of homogeneous bundles reduces to the study of G-bundles, i.e. to the study of representations of the subgroup P. More precisely, if $x \in G/P$ is the point associated to P, then the correspondence

$$(E, \varphi) \mapsto (E_x, \varphi|_P)$$

gives an equivalence between the category of G-bundles and the category of coherent algebraic representations of the group P, which will be denoted by P-mod. Let $\mathfrak{p}$ be the Lie algebra of the group P, $\mathfrak{n} \subset \mathfrak{p}$ the nilpotent radical and $\mathfrak{a} \subset \mathfrak{p}$ the Levi subalgebra, so that as a vector space $\mathfrak{p} = \mathfrak{a} \oplus \mathfrak{n}$. Also let $\mathfrak{g}$ be the Lie algebra of G, $\mathfrak{f} \subset \mathfrak{g}$ a Cartan subalgebra and $\mathfrak{b} \supset \mathfrak{f}$ a Borel subalgebra of $\mathfrak{g}$ such that $\mathfrak{p} \supset \mathfrak{b}$. Note that $\mathfrak{f}$ is also a Cartan subalgebra of the reductive algebra $\mathfrak{a}$. Finite dimensional representations of $\mathfrak{p}$ with integer weights are in one-one correspondence with finite dimensional algebraic representations of P, i.e. the corresonding categories are equivalent. Indeed, if we are given a representation of $\mathfrak{p}$, then restricting it to $\mathfrak{a}$ and integrating gives a representation of the corresponding Levi group $A \subset P$. Now, elements of $\mathfrak{n}$ are represented by nilpotent operators, because they raise the weight. Therefore their exponential is a sum of a finite number of generators and determines an extension of the representation to the whole group P.

Now consider an arbitrary representation M of the algebra $\mathfrak{p}$. When restricted to the reductive algebra $\mathfrak{a}$, it decomposes into a direct sum

$$\bigoplus_{\lambda \in \mathfrak{f}^*} M_\lambda \otimes V_\lambda.$$

Here λ runs over the integer weights which are dominant for the subalgebra $\mathfrak{a}$, M_λ is an irreducible representation of $\mathfrak{a}$ with highest weight λ and V_λ is the multiplicity space. We think of M_λ as the representation generated by the vector $|vac(\lambda)\rangle$ of weight λ, which is annihilated by the nilpotent subalgebra $\mathfrak{a} \cap \mathfrak{b}$. Here $V_\lambda = \mathrm{Hom}(M_\lambda, M)$.

For every root γ of the algebra $\mathfrak{g}$ with respect to the Cartan subalgebra $\mathfrak{f}$, we denote by e_γ the corresponding Chevalley generator of the algebra $\mathfrak{g}$. It is a root vector, i.e. $[h, e_\gamma] = \gamma(h) e_\gamma$. Let $N_{\gamma, \delta} \in \mathbf{Z}$ be the coefficients defined by the condition

$$[e_\gamma, e_\delta] = N_{\gamma, \delta}\, e_{\gamma + \delta}.$$

Let Δ be the root system of the algebra $\mathfrak{g}$ with respect to $\mathfrak{f}$, $\Delta_+ \subset \Delta$ a system of positive roots composed of roots of the subalgebra $\mathfrak{b}$. We will denote by $\Delta_+(P)$ the set of roots of $\mathfrak{n}$, and by $C^+(\mathfrak{a}) \subset \mathfrak{f}^*$ the set of dominant weights of the subalgebra $\mathfrak{a}$. Let $\alpha_1, \ldots, \alpha_n \in \Delta_+$ be a system of simple roots. The parabolic subalgebra is generated by $\mathfrak{b}$ and $e_{-\alpha_j}$ for $\alpha_j \in S$, where S is some subset of $\{\alpha_1, \ldots, \alpha_n\}$. The map $\mathfrak{n} \otimes M \to M$ given by the action of $\mathfrak{n}$ is a morphism of $\mathfrak{a}$-modules if we consider the adjoint action of $\mathfrak{a}$ on $\mathfrak{n}$. Decomposing $\mathfrak{n} \otimes M_\lambda$ into irreducible representations, we get a system of maps between the multiplicity spaces V_λ. There are situations in which such a decomposition is particularly simple. This is the case, first of all, when P is a Borel subgroup and, secondly, when all the simple components of G are in one of the series A,D,E [2, §6.4]. (These series are distinguished by the fact that the off-diagonal elements of the Cartan matrix are equal to 0 or -1.)

PROPOSITION 2. *In both the situations described above, there is an $\mathfrak{a}$-module isomorphism*

$$\mathfrak{n} \otimes M_\lambda \cong \bigoplus_{\substack{\gamma \in \Delta_+(P) \\ \lambda + \gamma \in C^+(\mathfrak{a})}} M_{\lambda+\gamma}$$

Every irreducible representation is taken with multiplicity one.

PROOF. In the case when P is a Borel subgroup, $\mathfrak{a} = \mathfrak{f}$ and the statement is clear. Consider the case when all the simple components of G lie in the series A,D,E. The character of the representation $\mathfrak{n}$ is the sum

$$\sum_{\gamma \in \Delta_+(P)} e^\gamma$$

and the character of M_λ is equal, by the Weyl formula, to

$$\frac{\sum_{w \in W(A)} \text{sgn}(w) e^{w(\lambda+\rho)}}{\sum_{w \in W(A)} \text{sgn}(w) e^{w(\rho)}}$$

Here $W(A)$ is the Weyl group of the reductive group A, ρ is half the sum of the positive roots of the algebra $\mathfrak{a}$. (By a positive root we mean here a root which lies in Δ_+.) Since $\Delta_+(P)$ is invariant under $W(A)$, we have

$$\left(\sum_{\gamma \in \Delta_+(P)} e^\gamma \right) \left(\sum_{w \in W(A)} \text{sgn}(w) e^{w(\lambda+\rho)} \right) = \sum_{\gamma \in \Delta_+(P)} \sum_{w \in W(A)} \text{sgn}(w) e^{w(\lambda+\gamma+\rho)}$$

It remains to check that the ratio of the right hand side to

$$\sum_{w \in W(A)} \text{sgn}(w) e^{w(\rho)}$$

is indeed a sum of characters of irreducible representations, because the weights $\lambda + \gamma$ may not be dominant. However, in our case, every nondominant weight of the form $\lambda + \gamma$, when increased by ρ, has a nontrivial stabiliser in $W(A)$. Therefore the corresponding summand in the alternating sum equals zero. We now prove this claim. We need to check that $\lambda + \gamma + \rho$ is a dominant weight, i.e. for any simple root α of the algebra $\mathfrak{a}$ we have $\langle \lambda + \gamma + \rho, \alpha^\vee \rangle \geqslant 0$, where $\alpha^\vee$ is the coroot corresponding to α [2, §6.1]. Since $\langle \rho, \alpha^\vee \rangle = 1$ and λ is dominant, it suffices to prove that $\langle \gamma, \alpha^\vee \rangle \geqslant -1$. Consider the intersection Δ_0 of the root system Δ with the two dimensional subspace generated by α and γ. It follows from our assumptions that this root system is isomorphic to A_2 or $A_1 \oplus A_1$. For both the systems we have mentioned, the statement we need follows immediately. This completes the proof.

Thus, when $\lambda + \gamma$ is dominant, there exist $\mathfrak{a}$-invariant projections $\pi_\gamma : \mathfrak{n} \otimes M_\lambda \to M_{\lambda+\gamma}$. We normalise them so that $e_\gamma \otimes |\, vac(\lambda)\, \rangle$ maps to $|\, vac(\lambda + \gamma)\, \rangle$. (Recall that e_γ is the Chevalley generator.) Let $i_\gamma : M_{\lambda+\gamma} \to \mathfrak{n} \otimes M_\lambda$ be the corresponding system of equivariant inclusions.

Returning to the P-module M, which decomposes as an $\mathfrak{a}$-module into a sum $\bigoplus M_\lambda \otimes V_\lambda$, we can use the action $\mathfrak{n} \otimes M \to M$ and the inclusions i_γ to obtain a system of linear maps $u_{\gamma,\lambda} : V_\lambda \to V_{\lambda+\gamma}$. More precisely, for $v \in V_\lambda$, we define $u_{\gamma,\lambda}(v)$ to be the element $e_\gamma(|\, vac(\lambda)\, \rangle) \otimes v)$ of the quotient of M by all isotropic components except $V_{\lambda+\gamma} \otimes M_{\lambda+\gamma}$. This element is the highest weight vector in the component $V_{\lambda+\gamma} \otimes M_{\lambda+\gamma}$ and so can be interpreted as an element of $V_{\lambda+\gamma}$.

PROPOSITION 3. *The maps $u_{\lambda,\gamma}$ satisfy the same commutation relations as the Chevalley generators, i.e.*

$$u_{\delta,\lambda+\gamma}\, u_{\gamma,\lambda} - u_{\gamma,\lambda+\delta}\, u_{\delta,\lambda} = N_{\gamma,\delta} u_{\gamma+\delta,\lambda} \qquad for \qquad \gamma, \delta \in \Delta_+(P).$$

PROOF. According to the definition, for $v \in V_\lambda$ we have

$$\left(u_{\delta,\lambda+\gamma}\, u_{\gamma,\lambda} - u_{\gamma,\lambda+\delta}\, u_{\delta,\lambda} \right) v = [e_\delta, e_\gamma]\left(|\, vac(\lambda) \otimes v\, \rangle \right)$$

with respect to the rest of the isotropic components. This gives the desired result, since M is an $\mathfrak{n}$-module.

Let us now consider the following preadditive category $\mathcal{A}(P)$. Its objects are the symbols (λ), for the integral, $\mathfrak{a}$-dominant weights $\lambda \in \mathfrak{f}^*$. Its morphisms are generated by the elementary morphisms $u_{\gamma,\lambda} : (\lambda) \to (\lambda + \gamma)$ for every $\gamma \in \Delta_+(P)$, subject to the same commutation relations that are satisfied by the e_γ. One can think of this category as a quiver with vertices (λ), arrows $u_{\gamma,\lambda}$ and the given relations.

THEOREM. *The category of finite-dimensional algebraic representations of the group P is equivalent, under the assumptions of Proposition 2, to the category of finite-dimensional $\mathcal{A}(P)$-modules, i.e. covariant functors from $\mathcal{A}(P)$ to the category of finite dimensional vector spaces, taking almost all objects to 0.*

PROOF. The correspondence in one direction has already been constructed. If, on the other hand, we are given the spaces V_λ and morphisms $u_{\gamma,\lambda}$ between them, then we have a morphism of $\mathfrak{a}$-modules

$$\mathfrak{n} \otimes \left(\bigoplus M_\lambda \otimes V_\lambda \right) \longrightarrow \bigoplus M_\lambda \otimes V_\lambda.$$

If the $u_{\gamma,\lambda}$ satisfy the required relations, then we get a $\mathfrak{p}(= \mathfrak{a} \ltimes \mathfrak{n})$-module structure on $\bigoplus M_\lambda \otimes V_\lambda$.

Now consider the case when X is the projective space $\mathbf{P}^n = \mathbf{P}(E)$. Then G is the general linear group $GL(n+1) = GL(E)$ and

$$P = \begin{pmatrix} GL(1) & W \\ 0 & GL(n) \end{pmatrix}$$

is the subgroup preserving a one-dimensional subspace $E_1 \subset E$. The nilpotent radical $\mathfrak{n}$ is the n-dimensional vector space $W = E/E_1$ with a trivial Lie algebra structure. The group $GL(1) = GL(E_1)$ acts on it by scalar multiplication and $GL(n) = GL(E/E_1)$ has the usual vector representation. The admissable weights in this case are $(\lambda_0, \lambda_1, \ldots, \lambda_n)$, where $\lambda_1 \geqslant \cdots \geqslant \lambda_n$ ($\lambda_i \in \mathbf{Z}$). For any space T, the irreducible representation of $GL(T)$ of highest weight α will be denoted by $\Sigma^\alpha T$. In this notation, the irreducible A-module M_λ is given by

$$M_\lambda = E_1^{\otimes \lambda_0} \otimes \Sigma^{(\lambda_1, \ldots, \lambda_n)} (E/E_1).$$

Corresponding to the representations E_1 and E/E_1, we have the bundles $\mathcal{O}(-1)$ and $T(-1)$ on $\mathbf{P}^n$.

The decomposition of $W \otimes M_\lambda$ into irreducible subspaces takes the form

$$W \otimes M_\lambda = \bigoplus_{i=1}^n M_{(\lambda_0 - 1, \lambda_1, \ldots, \lambda_i + 1, \lambda_{i+1}, \ldots, \lambda_n)}$$

(of course, on the right hand side we only consider the summands corresponding to dominant weights). The Chevalley generators e_i correspond to adding one to the ith coordinate ($1 \leqslant i \leqslant n$) and subtracting one from the 0th coordinate. Since the e_i commute with one another, the representations of $\mathcal{A}(P)$ are commutative diagrams of vector spaces V_λ. There is no interaction between the spaces V_λ and V_μ if $\sum_{i=0}^n \lambda_i \neq \sum_{i=0}^n \mu_i$. In other words, the category $\mathcal{A}(P)$ is the direct sum, over $m \in \mathbf{Z}$, of its full subcategories $\mathcal{A}_m(P)$, where

$$Ob\,\mathcal{A}_m(P) = \left\{ (\lambda) \mid \sum_{i=0}^n \lambda_i = m \right\}.$$

Twisting the representation M by the one-dimensional representation $\det(E)$ does not change the isomorphism class of the corresponding bundle on $\mathbf{P}^n$. Under this twisting, the V_λ change to $V_{(\lambda_0+1,\ldots,\lambda_n+1)}$. Therefore, modulo this twist there are $n+1$ "fibres" in the category $\mathcal{A}(P)$.

EXAMPLE. Consider the irreducible representation $\Sigma^\lambda E$ of the group $GL(E)$ as a representation of P. Here $\lambda_0 \geqslant \lambda_1 \geqslant \cdots \geqslant \lambda_n$. The restriction of this representation to $A = GL(1) \times GL(n)$ is $\bigoplus M_\mu$, where μ runs through all the collections $(\mu_0,\ldots,\mu_n)$ such that $\lambda_0 \geqslant \mu_1 \geqslant \lambda_1 \geqslant \mu_2 \geqslant \cdots \geqslant \mu_n \geqslant \lambda_n$ and $\sum_{i=0}^n \mu_i = \sum_{i=0}^n \lambda_i$ ([1]). In other words, the set of μ we get is an n-dimensional parallelepiped with edges of length $\lambda_i - \lambda_{i+1}$. The elementary morphisms in $\mathcal{A}(P)$ are parallel to these edges. In the given case they are isomorphic to one-dimensional vector spaces.

As is shown in [5], the representations $E_1^{\otimes i} \otimes \Sigma^\lambda E$ form an exceptional collection in the category P-mod, i.e. the endomorphisms of each object are the scalars and there are no higher Ext's between objects. Furthermore, these representations generate the derived category $D^b(P\text{-mod})$.

Finally, we will consider the case $X = \mathbf{P}^2$ and we will give examples of representations of the quiver $\mathcal{A}(P)$ which correspond to exceptional bundles on $\mathbf{P}^2$ of rank less than 100. There are seven such bundles, modulo twisting and conjugacy, and their ranks are 1, 2, 5, 13, 29, 34 and 89. The one-dimensional bundles are of course the $\mathcal{O}(i)$. The two dimensional ones are the twistings of the tangent bundle $\mathcal{T}_{\mathbf{P}^2}$. The five dimensional ones are equivalent, in the above sense, to the extension

$$0 \to \mathcal{T} \to E \to S^2\mathcal{T} \to 0,$$

which is the quotient of the space of differential operators of order less than or equal to two by the space of functions.

In the first two cases the corresponding representations of the quiver are concentrated at a point, i.e. the unique nontrivial vector space is one-dimensional and sits at the point $(-2a, a, a)$ (respectively $(-2a-1, a+1, a)$). In the third case the representation has the form of an arrow giving an isomorphism between two one dimensional spaces. The structure of the representations of $\mathcal{A}(P)$ in the remaining cases is depicted in the diagrams at the end of this paper. The numbers over the points denote the dimensions of the corresponding vector spaces. Points on the diagonal correspond to invertible sheaves and points on the kth straight line below and parallel to the diagonal correspond to twists of $S^k\mathcal{T}$. Only the arrows corresponding to nontrivial morphisms are drawn. Since exceptional bundles are indecomposable, we restrict ourselves in each diagram to one component of the category $\mathcal{A}(P)$. Twisting by invertible sheaves corresponds to translating parallel to the diagonal $x = y$ and dualising corresponds to reflecting in the line $x = -y$.

References

[1] BEILINSON, A.A., Coherent Sheaves on $\mathbf{P}^n$ and Problems in Linear Algebra, *Funk. An.*, **12** (1978) 68–69.

[2] BOURBAKI, N., *Lie Groups and Lie Algebras*, Chapters 4–6.

[3] CASSELS, J.W.S., *An Introduction to Diophantine Approximation*, C.U.P. (1957).

[4] GORODENTSEV, A.L., & RUDAKOV, A.N., Exceptional Vector Bundles on Projective Space, *Duke Math. J.*, **54** (1987) 115–130.

[5] KAPRANOV, M.M., On the Derived Categories of Coherent Sheaves on some Homogeneous Spaces, *Inv. Math*, **92** (1988) 479–508.

[6] ATIYAH, M.F., *K-Theory*, Benjamin (1963).

A.I. Bondal   M.M. Kapranov

Diagram 1.   Rank=13

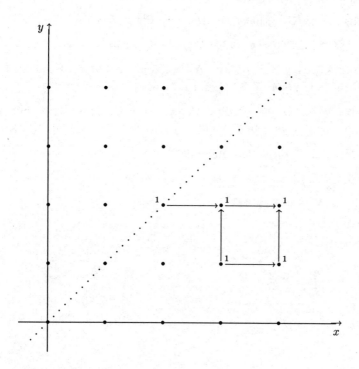

Diagram 2.   Rank=29

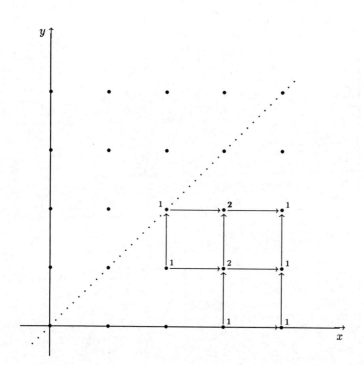

A.I. Bondal   M.M. Kapranov

Diagram 3.   Rank=34

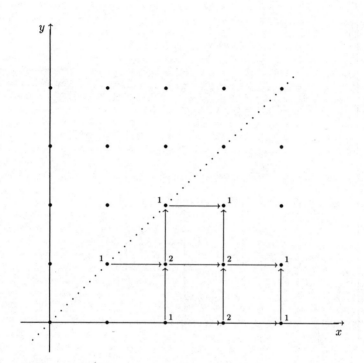

Diagram 4.   Rank=89

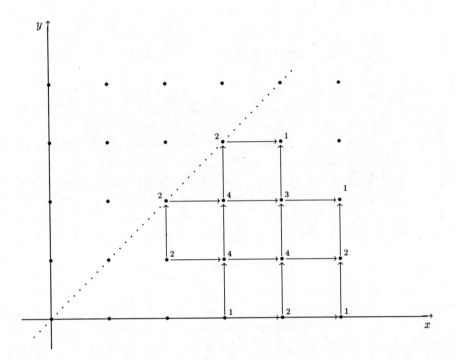

A.I. Bondal   M.M. Kapranov

7. Exceptional Objects and Mutations in Derived Categories

A. L. Gorodentsev

Introduction.

In this paper we develop the "Theory of Mutations" (or "Helix Theory") from an algebraic point of view. This theory arose as a convenient method of constructing and describing geometric objects such as exceptional sheaves on algebraic varieties.

First of all we will give a general definition of mutations which will work in an arbitrary category of coherent sheaves and makes it possible to overcome all the difficulties connected with the problem of "existence". After that we will study properties of mutations with respect to exceptional objects, mutations of exceptional pairs and so on; in other words, we obtain a generalization of all the technical results of [2]. As an example of the application of the techniques developed, we define *higher mutations* of sheaves which also always exist and as a consequence get a unified presentation of the theory of mutations on homogeneous surfaces. All this is contained in section 2.

Further on we deal with questions of the representation of the elements of the derived category by standard resolutions and our results parallel those of Kapranov [3], the only difference being that instead of using the language of B-resolvents we use the language of canonical mutations which leads to a more precise formulation of some results. The most important of which is that for every subcategory $T \subset D(X)$ generated by exceptional collections, we define the functors of left and right "orthogonal" projection $\langle T \mid$ and $\mid T \rangle$ onto the subcategory T, and functors L_T and R_T of "projections onto the left and right orthogonal complements" of T. This makes it possible to obtain left and right orthogonal decomposition of an arbitrary $Q \in \mathrm{Ob}(D(X))$, i.e. to build around Q canonical distinguished triangles with **RHom**-orthogonal edges:

$$(\langle T \mid Q \rangle, Q, L_T Q)$$
$$(R_T Q, Q, \langle Q \mid T \rangle).$$

We also show how to calculate the cohomology sheaves of the object $\langle T \mid Q \rangle$ using a Beilinson-type spectral sequence. The cohomological constructions which we need are recalled (usually without proofs) in section 1. In that section we also prove that on some surfaces all the exceptional objects of a derived category are given by the canonical images of exceptional sheaves and their shifts, i.e. studying exceptional sheaves on such surfaces on the level of derived categories does not distort their information content.

Two further important remarks.

In choosing the derived category as the object of study (and also in choosing the field $\mathbb{C}$ to be the base field) we are not conforming to the present trends but are adhering to geometric tradition. All the results can be reformulated for an arbitrary triangulated category, but instead of the complexes **R**Hom it is necessary to use the complexes Hom^i (quasi-isomorphic to them in this case) with trivial differentials. More detail can be found in the paper of A. I. Bondal in this collection.

On the other hand an important restriction is the necessity of existence in our category of exceptional objects and collections. In particular we develop a method not used before in the study of coherent sheaves on varieties with positive canonical class.

Finally I want to note that the results of section 3 were formulated as the result of a great number of talks with A. I. Bondal and M. M. Kapranov, and I am very grateful to them for these discussions.

1.   Basic Concepts.

We will recall here some basic contructions of homological algebra which will play an important rôle later. A more detailed presentation of the basic concepts related to derived categories can be found in the books [5] and [1].

Let X be an algebraic variety. As is customary, we will let Hot(X) and $D(X)$ denote the category of finite complexes and the bounded category of coherent sheaves on X, respectively. Similarly Hot$(\mathcal{V})$ and $D(\mathcal{V})$ will denote the corresponding categories for finite dimensional vector spaces. We will always assume that the base field is $\mathbb{C}$. δ will always denote the inclusion functor of the base category into the derived category, which takes an object to the complex whose only nonzero object is this object placed in the 0th position.

1.1.   The Functor RHom$(*, *)$.

The derived functor of the usual Hom functor on the category of coherent sheaves is the functor

$$\mathbf{R}\mathrm{Hom}(*, *) : D(X) \times D(X) \longrightarrow D(\mathcal{V}),$$

which on the level of the category Hot(X) assigns to a pair of complexes $A^\bullet$, $B^\bullet$, the total complex of the bicomplex Hom$(A^\bullet, B^\bullet)$:

$$\mathrm{Tot}\big(\mathrm{Hom}(A^\bullet, B^\bullet)\big)^i = \bigoplus_k \mathrm{Hom}_{\mathcal{O}_X}(A^k, B^{k+i}).$$

The following equality holds

$$H^i(\mathbf{R}\mathrm{Hom}(A,B)) = \mathrm{Hom}^i_{D(X)}(A,B),$$

so that for arbitrary coherent sheaves E and F on X

$$\mathrm{Ext}^i(E,F) = H^i(\mathbf{R}\mathrm{Hom}(\delta E, \delta F)).$$

If $A, B \in \mathrm{Ob}(D(X))$ and the filtrations

$$0 = F^0 A \to F^1 A \to \cdots \to F^{n-1} A \to F^n A = A$$
$$0 = F^0 B \to F^1 B \to \cdots \to F^{n-1} B \to F^n B = B$$

are given. Then in order to calculate $H^i(\mathbf{R}\mathrm{Hom}(A,B))$ there is a standard spectral sequence see [1],

$$E_1^{p,q} = \bigoplus_k H^{p+q}(\mathbf{R}\mathrm{Hom}(\mathrm{Gr}^i A, \mathrm{Gr}^{i-p} B)).$$

In particular, let $A = B$ and consider the canonical filtration of the complex

$$A: \quad 0 \longrightarrow A^0 \longrightarrow A^1 \longrightarrow \cdots \longrightarrow A^n \longrightarrow 0$$

in which $F^i A$ takes the form

$$0 \longrightarrow A^0 \longrightarrow A^1 \longrightarrow \cdots \longrightarrow A^{i-2} \longrightarrow \mathrm{Ker}(d_{i-1}) \longrightarrow 0$$

and the functor $\mathrm{Gr}^i(A): 0 \longrightarrow A^{i-2}/\mathrm{Ker}(d_{i-2}) \longrightarrow \mathrm{Ker}(d_{i-1}) \longrightarrow 0$ which is quasi-isomorphic to $\delta H^i(A)[-i]$. Then, from the equalities

$$H^{p+q}(\mathbf{R}\mathrm{Hom}(\delta H^i(A), \delta H^{i-p}(A)[p])) = \mathrm{Ext}^{2p+q}(H^i(A), H^{i-p}(A)),$$

we see that there exists a spectral sequence

$$E^{p,q} \Rightarrow H^{p+q}(\mathbf{R}\mathrm{Hom}(A,A))$$

for which

$$E_1^{p,q} = \mathrm{Ext}^{2p+q}(H^i(A), H^{i-p}(A)).$$

1.2. Exceptional Objects.

1.2.1. DEFINITION. An object A of $D(X)$ is *exceptional* if the canonical morphism

$$\gamma : \delta \mathbb{C} \longrightarrow \mathbf{R}\mathrm{Hom}(A, A) \tag{1}$$

(the image of which is the homotheties of the complex $A^\bullet$) is a quasi-isomorphism.

1.2.2. In general it is possible that there are no exceptional objects in $D(X)$. For example if X is a K3 surface then from Serre duality (see [4]) we have the isomorphism

$$H^0\big(\mathbf{R}\mathrm{Hom}(A, A)\big) \cong H^2\big(\mathbf{R}\mathrm{Hom}(A, A)\big),$$

the existence of which rules out the possibility that (1) is a quasi-isomorphism. It is possible, however, that on varieties with a negative canonical class such objects do exist and on projective spaces, quadrics and flag varieties such objects always exist— in particular, we have the images δE of exceptional sheaves E (i.e. sheaves for which $\mathrm{Hom}(E, E) = \mathbb{C}$, and $\mathrm{Ext}^i(E, E) = 0$). The proposition proved below shows that on some surfaces (in particular, on homogeneous surfaces) all exceptional objects of the derived category take the form $\delta E[i]$ for some exceptional sheaf E on X.

1.2.3. PROPOSITION. *Let X be an irreducible algebraic surface with $\dim H^0(-K_X) \geqslant 2$, which has no rigid torsion sheaves and let $A \in \mathrm{Ob}(D(X))$ be an exceptional object. Then exactly one of the cohomology sheaves $H^i(A)$ is nonzero.*

PROOF. It is easy to prove that from the condition $\dim H^0(-K_X) \geqslant 2$ it follows that for all torsion-free sheaves F and for all sheaves G

$$\begin{aligned} \text{either} \quad & \mathrm{Hom}(G, F) = \mathrm{Hom}(G, F(-K_X)) = 0 \\ \text{or} \quad & \dim \mathrm{Hom}(G, F) > \dim \mathrm{Hom}(G, F(-K_X)). \end{aligned} \tag{2}$$

Moreover, if there are no rigid torsion sheaves on the surface, then any rigid sheaf has to be torsion-free since any torsion subsheaf of a rigid sheaf on a surface is always rigid (see [4]). Combined with (2) this gives us, for arbitrary rigid sheaves on X, the following alternative

$$\begin{aligned} \text{either} \quad & \mathrm{Hom}(G, F) = \mathrm{Ext}^2(F, G) = 0, \\ \text{or} \quad & \dim \mathrm{Hom}(G, F) > \dim \mathrm{Ext}^2(F, G). \end{aligned}$$

Now let A be an exceptional object of $D(X)$. We consider the spectral sequence from §1.1. Its nonzero elements are concentrated in the range $0 \leqslant 2p + q \leqslant 2$, and in particular it is clear that $E_2 = E_\infty$ (see diagram)

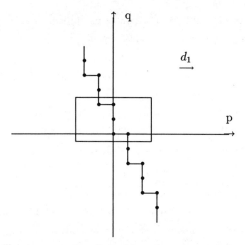

Let us consider in more detail the boxed area in the diagram:

$$\bigoplus \mathrm{Hom}(H^{i-1}, H^i) \xrightarrow{d_1^{-1,2}} \bigoplus \mathrm{Ext}^2(H^i, H^i) \longrightarrow 0$$

$$0 \longrightarrow \bigoplus \mathrm{Ext}^i(H^i, H^i) \longrightarrow 0$$

$$0 \longrightarrow \bigoplus \mathrm{Hom}(H^i, H^i) \xrightarrow{d_1^{0,1}} \bigoplus \mathrm{Ext}^2(H^i, H^{i-1})$$

Since $\bigoplus \mathrm{Ext}^1(H^1, H^1) = E_\infty^{0,1} \subset H^1(\mathbf{R}\mathrm{Hom}(A, A)) = 0$, all the sheaves $H^i = H^i(A)$ are rigid and therefore the following inequalities are satisfied

$$\dim \mathrm{Hom}(H^i, H^i) > \dim \mathrm{Ext}^2(H^i, H^i)$$
$$\dim \mathrm{Hom}(H^{i-1}, H^i) \geqslant \dim \mathrm{Ext}^2(H^i, H^{i-1}).$$

But this is possible if and only if only one of the H^i is different from zero because, as A is exceptional, the differential $d_1^{-1,2}$ would be an isomorphism and $d_1^{0,1}$ would be an epimorphism with one-dimensional kernel!

1.3. Complexes and their Associated Total Complexes.

By complex we always mean a sequence

$$0 \longrightarrow L^0 \longrightarrow L^1 \longrightarrow \cdots \longrightarrow L^n \longrightarrow 0,$$

where $L^i \in \mathrm{Ob}(D(X))$, n is a large enough fixed number and the morphisms ($d^i : L^i \to L^{i+1}$) $\in \mathrm{Mor}(D(X))$ satisfy $d^{i+1} d^i = 0$.

1.3.1. DEFINITION. A *right Postnikov system* (see [3]) of a complex $L^\bullet$ is a diagram of the form

$$
\begin{array}{ccccccccccc}
L^0 & & L^1[-1] & & & & L^{n-2}[-n+2] & & L^{n-1}[-n+1] & & L^n[n] \\
{\scriptstyle j_{-1}}\nearrow & {\scriptstyle i_0}\searrow & {\scriptstyle j_0}\nearrow & {\scriptstyle i_1}\searrow & {\scriptstyle j_1}\nearrow & \cdots & \nearrow{\scriptstyle j_{n-2}} & \searrow{\scriptstyle i_{n-1}} & \nearrow{\scriptstyle j_{n-1}} & \searrow{\scriptstyle d_{n-1}} & \nearrow{\scriptstyle id} \\
B^0 & \leftarrow & B^1 & \leftarrow & B^2 \leftarrow & & \leftarrow & & B^{n-1} & \leftarrow & L^n[-n+1]
\end{array}
$$

in which all the distinguished triangles, horizontal morphisms and morphisms i_k have degree 0, the morphisms j_k have degree 1, and the relation $j_k i_k = d^k$ is satisfied.

1.3.2. Right Postnikov sytems are not uniquely determined and do not always exist. If they do exist for a complex $L^\bullet$ then this complex is called *regular* and all possible objects B^0 appearing in the lower left hand corners of different Postnikov systems are called *right twists* and the set of them will be denoted by $\mathrm{Tot}(L^\bullet)$. For more details about this (and also an equivalent notion of left twist and regularity) see [3].

1.3.3. An important example of a twist is furnished by the standard realisation of a filtered object

$$A = F^0 \leftarrow F^1 \leftarrow \cdots \leftarrow F^n \leftarrow 0$$

in the form of a twist of the factors in the filtration (see [1]).Therefore the regular complex can be considered as a generalization of the notion of a filtered object. In particular, by the very same argument as was used in the case of a filtered object one can establish (see [3]) for an arbitrary cohomological functor $\Phi^\bullet$, the existence of the spectral sequence $E^{p,q} \Rightarrow \Phi^{p+q}(B^0)$ with $E_1^{p,q} = \Phi^q(L^p)$ for an arbitrary $B^0 \in \mathrm{Tot}(L^\bullet)$.

1.4. Orthogonal decomposition of Functors.

Further on we shall be constantly applying the following method of constructing functors on a triangulated category T, used in [3].

1.4.1 DEFINITION. We call an ordered pair of functors (Ψ_1, Ψ_2), $\Psi_1, \Psi_2 : T \to T$ *orthogonal* if for all $Q \in \mathrm{Ob}(T)$

$$\mathrm{Hom}^i(\Psi_1(Q), \Psi_2(Q)) = 0 \qquad \text{where } i = -1, 0.$$

An *orthogonal decomposition* of a functor $\Phi : T \to T$ is any pair of orthogonal functors (Ψ_1, Ψ_2) together with natural transformations

$$\lambda : \Psi_1 \longrightarrow \Phi$$
$$\rho : \Phi \longrightarrow \Psi_2$$
$$\omega : \Phi_2 \longrightarrow \Psi_1[1]$$

which for any $Q \in \mathrm{Ob}(T)$ induces a distinguished triangle

$$\big(\Psi_1(Q), \Phi(Q), \Psi_2(Q)\big),$$

which will be called the *canonical triangle* associated to the given orthogonal decomposition.

PROPOSITION.

(1) Let (Ψ_1, Ψ_2) be an orthogonal pair of functors. Then any natural transformation $\omega : \Psi_2 \to \Psi_1[1]$ can be prolonged in a natural way to an orthogonal decomposition of some functor $\Phi : T \to T$, which is unique up to a natural isomorphism of functors.

(2) Let Ψ_1, Φ be an arbitrary pair of functors and $\lambda : \Psi_1 \to \Phi$ be a natural transformation such that for all $Q \in \mathrm{Ob}(T)$ and any cone Z over the morphism $\lambda(Q)$, $\mathrm{Hom}^i(\Psi_1(Q), Z) = 0$ for $i = -1, 0$. Then λ can be prolonged in a natural way to an orthogonal decomposition (Ψ_1, Ψ_2) of the functor Φ, in which Ψ_2 is uniquely defined up to a natural isomorphism of functors.

(3) Let Φ, Ψ_2 be an arbitrary pair of functors and $\rho : \Psi_2 \to \Phi$ be a natural transformation such that for any $Q \in \mathrm{Ob}(T)$ and any cone Z over the morphism $\rho(Q)$, $\mathrm{Hom}^i(Z, \Psi_2(Q)) = 0$ for $i = 0, +1$. Then ρ can be prolonged in a natural way to an orthogonal decomposition (Ψ_1, Ψ_2) of the functor Φ, in which Ψ_1 is uniquely defined up to a natural isomorphism of functors.

The proofs of all three statements are identical and are based on the existence of canonical cones over the natural transformations given in the hypotheses. Let us suppose that we are in situation (1). Then for any $Q \in \mathrm{Ob}(T)$ and arbitrary distinguished triangle

$$(\Psi_1(Q), Z, \Psi_2(Q))$$

induced by the morphism $\omega(Q) : \Psi_2(Q) \to \Psi_1(Q)[1]$, the two other morphisms $\Phi_1(Q) \to Z$ and $Z \to \Phi_2(Q)$ are determined in a canonical way as images of the identity morphisms $\mathrm{id}_{\Psi_1(Q)}$ and $\mathrm{id}_{\Psi_2(Q)}$ under the identifications

$$\mathrm{Hom}^0(\Psi_1(Q), \Psi_2(Q)) \simeq \mathrm{Hom}(\Psi_1(Q), Z)$$
$$\mathrm{Hom}^0(\Psi_1(Q), \Psi_2(Q)) \simeq \mathrm{Hom}^0(Z, \Psi_2(Q))$$

obtained by applying the functors $\mathrm{Hom}(\Psi_1(Q), *)$ and $\mathrm{Hom}(*, \Psi_2(Q))$ to our triangle. This implies that if

$$(\Psi_1(Q), Z', \Psi_2(Q))$$

is a second distinguished triangle which prolongs the morphism $\omega(Q)$, then there is a unique isomorphism of this triangle to the earlier one, which is the identity on the boundaries. This proves that the equivalence class of the functors Φ which we were looking for exists and is unique.

2.   Mutations of Pairs.

2.1. Canonical Morphisms.

Let us consider the functor

$$(V, A) \mapsto V \otimes A : D(\mathcal{V}) \times D(X) \longrightarrow D(X),$$

which assigns to a complex of sheaves and a complex of vector spaces, the tensor product of the complex of sheaves with the complex of sections of the trivial vector bundles corresponding to the vector spaces by fixing any bases. This functor amounts to taking direct sums of sheaves and therefore is a bi-additive exact functor.

There is a canonical morphism

$$lcan : \mathbf{R}\mathrm{Hom}(A, B) \otimes A \longrightarrow B \tag{3}$$

which, on the level of the category Hot, is obtained as the composition of the projection and the canonical map in the following diagram

$$
\begin{array}{ccc}
\left[\mathrm{Tot}\big(\mathrm{Hom}(A^{\bullet}, B^{\bullet})\big) \otimes A \right]^{n} & \overset{lcan}{\dashrightarrow} & B^{n} \\[2mm]
\| \mathrm{def} & & \uparrow \\[2mm]
\displaystyle\bigoplus_{i+j=n} \left(\bigoplus_{k} \mathrm{Hom}(A^{k}, B^{k+i}) \otimes A^{j} \right) & \longrightarrow & \displaystyle\bigoplus_{i+j=n} \mathrm{Hom}(A^{j}, B^{j+i}) \otimes A^{j}.
\end{array}
$$

This construction can be carried out in a natural way in the derived category, where, as was mentioned in the introduction, instead of the complex $\mathbf{R}\mathrm{Hom}$ we can use the $D(\mathcal{V})$-isomorphic complex composed of $\mathrm{Hom}^{i}_{D(X)}(A, B) = H^{i}(\mathbf{R}\mathrm{Hom}(A, B))$ with zero differentials and the result will be the same. In this form our construction can be applied to an arbitrary triangulated category; nevertheless, we shall confine ourselves to the notation introduced earlier. The morphism (3) could also be defined in an abstract way by saying that it is induced by an exact transformation under the identification

$$\mathbf{R}\mathrm{Hom}\big(\mathbf{R}\mathrm{Hom}(A, B) \otimes A, B\big) \simeq \mathbf{R}\mathrm{Hom}(A, B)^{*} \otimes \mathbf{R}\mathrm{Hom}(A, B).$$

Since the right hand side is also isomorphic to

$$\mathbf{R}\mathrm{Hom}\big(A, \mathbf{R}\mathrm{Hom}(A, B)^{*} \otimes B\big)$$

we have, dually to (3), a canonical morphism

$$rcan : A \longrightarrow \mathbf{R}\mathrm{Hom}(A, B)^{*} \otimes B$$

(we assume everywhere that the complex $V^{\bullet}$ dual to the complex $V^{*\bullet}$ has components $(V^{*})^{i} = (V^{-i})^{*}$). Interchanging A and B we can rewrite the last transformation for the pair (B, A)

$$rcan : B \longrightarrow \mathbf{R}\mathrm{Hom}(B, A)^{*} \otimes A. \tag{4}$$

2.2. Definition of Mutations.

The canonical morphisms (3) and (4) define natural transformations of functors

$$lcan : \mathbf{R}\mathrm{Hom}(A, *) \otimes A \longrightarrow Id_D(X)$$
$$rcan : Id_D(X) \longrightarrow \mathbf{R}\mathrm{Hom}(*, A)^* \otimes A$$

so that the objects $L_A B$ and $R_A B$ are defined (in general, they are only defined up to a noncanonical isomorphism). These objects are called the *left and right shifts* (or *mutations*) of B via A and can be inserted into the distinguished triangles

$$\left(\mathbf{R}\mathrm{Hom}(A, B) \otimes A, B, L_A B\right) \tag{5}$$
$$\left(R_A B, B, \mathbf{R}\mathrm{Hom}(B, A)^* \otimes A\right). \tag{6}$$

If A is an exceptional object of $D(X)$, then applying $\mathbf{R}\mathrm{Hom}(A, *)$ and $\mathbf{R}\mathrm{Hom}(*, A)$ to (5) and (6), one can convince oneself that conditions (2) and (3) of Proposition 1.4.2 are satisfied. Therefore for any exceptional object A, L_A and R_A will be exact additive functors, uniquely determined up to a natural isomorphism. By an abuse of notation we follow the terminology of [2] and call triangles (5) and (6) *canonical*.

2.3. Properties of Mutations.

According to the above, R_A and L_A behave well in the case when A is an exceptional object of $D(X)$.

2.3.1 DEFINITION. Let A be an exceptional object and B be any object of $D(X)$. Then $\mathbf{R}\mathrm{Hom}(A, L_A B) \simeq 0$. If we additionally assume that $\mathbf{R}\mathrm{Hom}(B, A) \simeq 0$, then the following properties will hold

(1) There is a isomorphism, natural with respect to B,

$$\mathbf{R}\mathrm{Hom}(L_A B, A)^* \simeq \mathbf{R}\mathrm{Hom}(A, B)[1],$$

which induces an isomorphism of triangle (5) with the twisted triangle

$$\left(R_A L_A B, L_A B, \mathbf{R}\mathrm{Hom}(L_A B, A)^* \otimes A\right)$$

so that $R_A L_A B = B$.

(2) For any $C \in \mathrm{Ob}(D(X))$ there is an isomorphism

$$\mathbf{R}\mathrm{Hom}(B, C) \simeq \mathbf{R}\mathrm{Hom}(L_A B, L_A C),$$

natural with respect to B and C, which induces an isomorphism between the triangle

$$\left(\mathbf{R}\mathrm{Hom}(L_A B, L_A C) \otimes L_A B, L_A C, L_{L_A B} L_A C\right)$$

and the triangle

$$L_A\left[\left(\mathbf{R}\mathrm{Hom}(B, C) \otimes B, C, L_B C\right)\right] = \left(\mathbf{R}\mathrm{Hom}(B, C) \otimes L_A B, L_A C, L_A L_B C\right)$$

so that $L_{L_A B} L_A C \simeq L_A L_B C$.

PROOF. The very first statement can be proved by applying $\mathbf{R}\mathrm{Hom}(A,*)$ to the triangle (5). In order to obtain isomorphism (1) we must apply $\mathbf{R}\mathrm{Hom}(*,A)$ to (5). In order to prove (2) we first apply $\mathbf{R}\mathrm{Hom}(B,*)$ to the triangle

$$\left(\mathbf{R}\mathrm{Hom}(A,C)\otimes A, C, L_A C\right)$$

to obtain the natural isomorphism $\mathbf{R}\mathrm{Hom}(B,C)\simeq\mathbf{R}\mathrm{Hom}(B,L_A C)$, and then apply $\mathbf{R}\mathrm{Hom}(*,L_A C)$ to (5) to obtain the second natural isomorphism $\mathbf{R}\mathrm{Hom}(B,L_A C)\simeq\mathbf{R}\mathrm{Hom}(L_A B,L_A C)$.

2.3.2 *The Duality Principle.* This says that *any theorem about left mutations can be converted into a theorem about right mutations using the formal substitution:*

$$R_A B = \left(L_{A^*} B^*\right)^*.$$

This can be explained by the fact that a formal application of the $*$ operator to the triangle (6) gives us a triangle of the form (5) for the pair A^*, B^*.

Henceforth we shall leave it to the reader to formulate and prove properties of right mutations dual to the properties of left mutations which we shall prove. This is entirely mechanical. For example, from Proposition 2.3.1 we get:

PROPOSITION. *Let A be an exceptional object and B any object of $D(X)$. Then $\mathbf{R}\mathrm{Hom}(R_A B, A) = 0$. If, in addition, we assume that $\mathbf{R}\mathrm{Hom}(A, B) = 0$, then*

(1) there is an isomorphism $\mathbf{R}\mathrm{Hom}(A, R_A B) \simeq \mathbf{R}\mathrm{Hom}(B, A)^*[-1]$, natural with respect to B, which identifies the right shift of B with the left shift of $R_A B$.

(2) For any $C \in \mathrm{Ob}(D(X))$ there is an isomorphism

$$\mathbf{R}\mathrm{Hom}(C, B) \simeq \mathbf{R}\mathrm{Hom}(R_A C, R_A B),$$

natural with respect to B and C identifying the canonical triangles

$$\left(R_{R_A B}R_A C, R_A C, \mathbf{R}\mathrm{Hom}(R_A C, R_A B)^* \otimes R_A B\right)$$

and

$$\left(R_A R_B C, R_C, \mathbf{R}\mathrm{Hom}(C, B)^* \otimes R_A B\right),$$

which is obtained by applying the functor R_b to the canonical triangle

$$\left(R_A C, C, \mathbf{R}\mathrm{Hom}(C, A)^* \otimes A\right)$$

so that $R_B R_A C = R_{R_B A} R_B C$.

2.3.4. Combining 2.3.4 with 2.3.1 we obtain

COROLLARY. *If A is an exceptional object of $D(X)$, then*

$$L_A \circ R_A \circ L_A = L_A$$
$$R_A \circ L_A \circ R_A = R_A.$$

2.4. Exceptional Pairs.

The ordered pair (A, B) of exceptional objects of $D(X)$ is called *exceptional* if $\mathbf{R}\mathrm{Hom}(B, A) \simeq 0$. The *left* and *right mutations* of a pair (A, B) are, respectively,

$$L(A, B) = (L_A B, A)$$
$$R(A, B) = (B, R_A B).$$

From the previous sections, it follows immediately that mutations of exceptional pairs are again exceptional.

Exceptional pairs are used in the construction of exceptional sheaves on varieties. It is then useful to bear in mind the following relations

$$L_{A[1]}B = L_A B$$
$$L_A(B[1]) = (L_A B)[1] \tag{7}$$

which can be easily obtained from the obvious equalities

$$V[k] \otimes A[m] = (V \otimes A)[k + m]$$
$$\mathbf{R}\mathrm{Hom}(A[k], B[m]) = \mathbf{R}\mathrm{Hom}(A, B)[m - k].$$

Dually for right mutations.

2.5. Mutations of Sheaves.

For any two coherent sheaves E and F on X, we will define the *left mutation* of F via E by putting

$$\lambda_E^i F = H^i(L_{\delta E}\delta F).$$

In this way left mutations always exist and are uniquely determined.

The long exact sequence in cohomology of triangle (5) in this case leads to the sequence

$$0 \to \lambda_E^{-1}F \to \mathrm{Hom}(E, F) \otimes E \to F \to \lambda_E^0 F \to \mathrm{Ext}^1(E, F) \otimes E \to 0 \tag{8}$$

and also to the relations $\lambda_E^{i-1}F = \mathrm{Ext}^i(E, F) \otimes E$ for $i \geqslant 1$. The question of which mutations of the form $\lambda_E^i F$ are different from zero is subtle. It is, however, sometimes possible to show (see also the papers of Bondal and Karpov in this collection) that only one left mutation is nonzero. Such pairs (E, F) are called *left regular*. In particular the left mutation 2.4 for left regular pairs, which is automatically exceptional pair, is well defined.

2.5.1 Example. Let X be a surface satisfying the conditions of 1.2.3, then passing to the derived category does not add new exceptional objects, other than the twisted images of exceptional sheaves. Consequently, all the exceptional pairs of sheaves are regular. If (E, F) is such a pair, then from the condition $\mathrm{Hom}(F, E) = 0$ which holds because of properties of the surface X, it follows that $\mathrm{Ext}^2(E, F) = 0$, so only one of the two mutations $\lambda_E^{-1} F$, $\lambda_E^0 F$ can be nonzero. Calculating the cohomology sheaves in sequence (8), taking into account the properties of our surface, shows that this nonzero mutation has to be represented by a short exact sequence, i.e. $\mathrm{Ext}^i(E, F)$ is nonzero either just for $i = 0$ or just for $i = 1$. In this way, if we denote this unique nonzero Ext group by $W(E, F)$ (the equality $W(E, F) = 0$ means that we have full orthogonality of E and F), and the unique nonzero mutation by $L_E F$, then we can formulate the following

COROLLARY. *All the exceptional pairs (E, F) on the surface X are left regular, the left mutation can be inserted into one of the short exact sequences:*

$$\text{(``division'')} \qquad 0 \longrightarrow L_E F \longrightarrow \mathrm{Hom}(E, F) \otimes E \longrightarrow F \longrightarrow 0$$

$$\text{(``recoil'')} \qquad 0 \longrightarrow \mathrm{Hom}(E, F) \otimes E \longrightarrow F \longrightarrow L_E F \longrightarrow 0 \qquad (9)$$

$$\text{(``extension'')} \qquad 0 \longrightarrow F \longrightarrow L_E F \longrightarrow \mathrm{Ext}^1(E, F) \otimes E \longrightarrow 0$$

and $W(L_E F, E) \simeq W(E, F)^*$.

We leave it to the reader to give the definition of a right mutation and to prove for these the analogues of all the properties formulated above.

3. Exceptional Collections and Orthogonal Decompositions.

3.1. Definition.

The ordered collection of objects $\varepsilon = (A_1, A_2, \ldots, A_k) \in \mathrm{Ob}(D(X))^k$ is called *exceptional* if each A_i is exceptional and $\mathbf{R}\mathrm{Hom}(A_i, A_j) = 0$ for $i > j$.

3.1.1. As usual, by mutations of exceptional collections we will mean mutations of any pair of neighbouring elements in the form defined in 2.4. One can check immediately that mutations of exceptional collections are again exceptional collections.

3.1.2. To each exceptional collection ε we will associate an exact triangulated subcategory $Tr(\varepsilon) \subset D(X)$, generated by ε, i.e. the set of all the objects which can be obtained by use of elements from ε, tensoring with elements of $D(\mathcal{V})$ and forming cones of all possible morphisms. It is clear that if ε and ε' are obtained from one another by mutations then $Tr(\varepsilon) = Tr(\varepsilon')$.

3.1.3. DEFINITION. We call any subcategory $T \subset D(X)$ which can be represented in the form $T = Tr(\varepsilon)$ for some exceptional collection ε *admissible* and we call ε a *normal basis* of T in this case. A category T possessing a normal basis composed of k elements will also be called a *k-subcategory*.

3.2. Projections.

The basic result of this section will be the following

PROPOSITION. *Let $T \subset D(X)$ be an admissible subcategory, then there exist exact additive functors*

$$Q \longmapsto \langle T \mid Q \rangle : D(X) \longrightarrow T$$
$$Q \longmapsto \langle Q \mid T \rangle : D(X) \longrightarrow T$$

which act as the identity on the subcategory T, and which are called left *and* right projections *onto T, such that for any $Q \in \mathrm{Ob}(D(X))$*

$$\mathbf{R}\mathrm{Hom}(A, Q) = \mathbf{R}\mathrm{Hom}(A, \langle T \mid Q \rangle)$$
$$\mathbf{R}\mathrm{Hom}(Q, A) = \mathbf{R}\mathrm{Hom}(\langle Q \mid T \rangle, A)$$

$$(10)$$

for all possible $A \in \mathrm{Ob}(T)$).

(In this way, $\langle T \mid$ and $\mid T \rangle$ are right and left adjoint functors to the inclusion functor $i_T : T \to D(X)$, and hence are uniquely determined.)

We will prove this proposition in 3.2.5, but for now we will list some of its immediate consequences generalizing the construction of mutations from §2.

3.2.2. Canonical morphisms. If we put $A = \langle T \mid Q \rangle$ and $A = \langle Q \mid T \rangle$ respectively in (10) we get:

$$\mathbf{R}\mathrm{Hom}(\langle T \mid Q \rangle, Q) \simeq \mathbf{R}\mathrm{Hom}(\langle T \mid Q \rangle, \langle T \mid Q \rangle)$$
$$\mathbf{R}\mathrm{Hom}(Q, \langle Q \mid T \rangle) \simeq \mathbf{R}\mathrm{Hom}(\langle Q \mid T \rangle, \langle Q \mid T \rangle)$$

and the identity morphisms on the right hand side induce canonical morphisms

$$rcan : Q \longrightarrow \langle Q \mid T \rangle$$
$$lcan : \langle T \mid Q \rangle \longrightarrow Q$$

$$(11)$$

and we obtain

COROLLARY. *There is a natural transformation of functors*

$$i_T \circ \langle T \mid \xrightarrow{lcan} \mathrm{Id}_{D(X)} \xrightarrow{rcan} i_T \circ \mid T \rangle.$$

3.2.3. Shifts. The *left* and *right shift* of $Q \in \mathrm{Ob}(D(X))$ with respect to an admissible category T are objects $L_T Q$ and $R_T Q$ which extend the canonical morphisms in (11) to the distinguished triangles

$$(\langle T \mid Q \rangle, Q, L_T Q)$$
$$(R_T Q, Q, \langle Q \mid T \rangle).$$

$$(12)$$

It is clear that, in a similar way to §2, L_T and R_T will be exact additive functors.

3.2.4 Orthogonal decomposition. Applying the functors $\mathbf{R}\mathrm{Hom}(A, *)$ and $\mathbf{R}\mathrm{Hom}(*, A)$ to (12) for any $A \in \mathrm{Ob}(D(X))$ we get, as in §2,

COROLLARY. $\mathbf{R}\mathrm{Hom}(A, L_T Q) \simeq \mathbf{R}\mathrm{Hom}(R_T Q, A) \simeq 0$ for any $A \in \mathrm{Ob}(T)$ and $Q \in \mathrm{Ob}(D(X))$.

In this way, the pairs of functors $\langle T \mid, L_T$ and $R_T, \mid T \rangle$ determine an orthogonal decomposition of $D(X)$ and L_T and R_T can be considered as "projections onto the right and left orthogonal complements" to T in $D(X)$.

3.2.5 Construction of Projections. Firstly we will construct the left projection $\langle T \mid$ by induction on the length of a normal basis of T. For admissible 1-subcategories $T = Tr(A)$ the projection already exists, in particular, as constructed in §2, we have

$$\langle T \mid = \mathbf{R}\mathrm{Hom}(A, *) \otimes A.$$

It is clear that, in this case, all the facts listed above can be converted to properties of mutations which we already know from §2.

Suppose that we have already constructed the projections to all admissible k-subcategories (or, in other words, we have properties 3.2.2–3.2.4 for these subcategories), and let $T = Tr(A_1, \ldots, A_k, B)$ be an admissible $(k + 1)$-subcategory. We put $S = Tr(A_1, \ldots, A_k)$.

It is obvious that the functors $\langle S \mid$ and $\mathbf{R}\mathrm{Hom}(B, *) \otimes L_S B$ generate an orthogonal pair. We will define a natural transformation of the form 1.4.2(1) between them taking the composition of natural transformations

$$\mathbf{R}\mathrm{Hom}(B, *) \otimes L_S B \longrightarrow \mathbf{R}\mathrm{Hom}(B, *) \otimes \langle S \mid [1]$$
$$\mathbf{R}\mathrm{Hom}(B, *) \otimes \langle S \mid \longrightarrow \langle S \mid,$$

where the first one is obtained from the induction hypothesis by taking the tensor product of the first triangle in (12) with $\mathbf{R}\mathrm{Hom}(B, *)$ (in (12) we need to take $T = S$!), and the second is obtained by applying the functor $\langle S \mid$ to the natural transformation $lcan$ in §2 (take $A = B$ in 2.2). It now follows from 1.4.2 that the canonical trangles

$$(\langle S \mid Q \rangle, \langle T \mid Q \rangle, \mathbf{R}\mathrm{Hom}(B, Q) \otimes L_S B) \tag{13}$$

correctly determine the class of isomorphic functors $\langle T \mid$.

In order to check whether $\langle T \mid$ has the desired property (10), it is enough to substitute generators of our category for A. Since $\mathbf{R}\mathrm{Hom}(A_i, L_S B) = 0$, if we apply the functor $\mathbf{R}\mathrm{Hom}(A_i, *)$ to (13), the induction hypothesis implies that we get the following series of isomorphisms

$$\mathbf{R}\mathrm{Hom}(A_i, \langle T \mid Q \rangle) \simeq \mathbf{R}\mathrm{Hom}(A_i, \langle S \mid Q \rangle) \simeq \mathbf{R}\mathrm{Hom}(A, Q).$$

Then, by hypothesis we have that $\mathbf{R}\mathrm{Hom}(B, \langle S \mid Q \rangle) = 0$ (because $\langle S \mid Q \rangle \in \mathrm{Ob}(S)$), and moreover we have that $\mathbf{R}\mathrm{Hom}(B, L_S B) \simeq \varepsilon C$, which is proved by applying $\mathbf{R}\mathrm{Hom}(B, *)$ to the canonical triangle

$$(\langle S \mid B \rangle, B, L_S B).$$

Consequently, if we apply $\mathbf{R}\mathrm{Hom}(B, *)$ to (13), we get an isomorphism

$$\mathbf{R}\mathrm{Hom}(B, \langle T \mid Q \rangle) \simeq \mathbf{R}\mathrm{Hom}(B, Q).$$

In this way, the functor $\langle T \mid$ is constructed and its properties have been established. To construct $\mid T \rangle$ one can either apply the duality principle 2.3.2 or represent T in the form $Tr(B, A_1, \ldots, A_k)$ and put $S = Tr(A_1, \ldots, A_k)$ and then construct the triangle

$$\left(\mathbf{R}\mathrm{Hom}(Q, B)^* \otimes R_S B, \langle Q \mid T \rangle, \langle Q \mid S \rangle\right)$$

dual to (13). Since the argument is exactly analogous to the one given above, we leave the details to the reader.

3.2.6. The properties of shifts with respect to an admissible category are analogous to properties of ordinary mutations proved in §2. For example, applying exactly the same reasoning as in 2.3.1, we deduce from relations 3.2.1 and 3.2.4.

COROLLARY. *Let T, S be an orthogonal pair of admissible subcategories (i.e. $\mathbf{R}\mathrm{Hom}(B, A) = 0$ for any $B \in \mathrm{Ob}(T)$, $A \in \mathrm{Ob}(S)$). Then $L_S(T)$ is an admissible subcategory and*

(1) for any $B \in \mathrm{Ob}(T)$ $\langle S \mid B \rangle = \langle L_S B \mid S \rangle[-1]$ so that $R_S \circ L_S|_T = Id_T$;

(2) for any $C \in \mathrm{Ob}(D(X))$

$$\langle L_S(T) \mid L_S C \rangle = \langle T \mid C \rangle$$

so that

$$L_S \circ L_T = L_{L_S(T)} \circ L_S.$$

The formulation of the analogues of 2.3.2–2.3.4 and their proofs are easy and can safely be left to the reader.

COROLLARY. *Let $(A_1, A_2, \ldots, A_k)$ be a normal basis of the admissible subcategory $T \subset D(X)$. Then for any $Q \in \mathrm{Ob}(D(X))$*

$$L_T Q = L_{A_1} L_{A_2} \cdots L_{A_k} Q \tag{14}$$

(and, in particular, the right hand side of this equation does not depend on the choice of the normal basis for T).

3.3. Calculating Projections.

Let $T = Tr(A_0, \ldots, A_k)$ be an admissible $(k+1)$-subcategory in $D(X)$. We compute the cohomology groups $H^i(\langle T \mid Q \rangle)$ in terms of $\mathrm{Hom}^r_{D(X)}(A_p, Q) \otimes H^s(L^{(p)} A_p)$, where $L^{(p)} A_p$, as usual, denotes the p-fold shift of A_p in the collection $(A_0, A_1, \ldots, A_k)$,

i.e. $\qquad L^{(p)} A_p = L_{A_0} L_{A_1} \ldots L_{A_{p-1}} A_p = L_{Tr(A_0, \ldots, A_{p-1})} A_p.$

From now on it will be convenient to denote the subcategory $Tr(A_0, \ldots, A_{i-1})$ by T_i, and the objects $\left[\mathbf{R}\mathrm{Hom}(A_{k-i}, Q \otimes L^{(k-i)} A_{k-i} \right][i]$ by $\mathcal{L}^i(Q)$.

PROPOSITION. *There exists a regular complex defined over the category T:*

$$0 \longrightarrow \mathcal{L}^0(Q) \longrightarrow \mathcal{L}^1(Q) \longrightarrow \cdots \longrightarrow \mathcal{L}^k(Q) \longrightarrow 0$$

which has a canonical right twist isomorphic to $\langle T \mid Q \rangle$.

PROOF. We will use the inductive construction of the functor $\langle T \mid$, presented in 3.2.5, using an increasing filtration of the category T by the categories T_i. The projections $\langle T_i \mid$ can be inserted into the series of distinguished triangles

$$(\langle T_i \mid Q \rangle, \langle T_{i+1} \mid Q \rangle, \mathcal{L}^{k-i}(Q)[-k+1])$$

of the form in (13), and combining them we also get a right Postnikov system for the complex $\mathcal{L}^\bullet(Q)$.

3.3.2. Recalling what was said in 1.3.3 we get the following

COROLLARY. *There is a spectral sequence $E^{p,q} \Rightarrow H^{p+q}(\langle T \mid Q \rangle)$, in which the E_1 term takes the form*

$$E_1^{p,q} = \bigoplus_{r+s=q} \mathrm{Hom}^r_{D(X)}(A_{k-i}, Q) \otimes H^s(L^{(k-i)} A_{k-i}).$$

Similar representations can also be obtained for the right projection $\langle Q \mid T \rangle$.

3.4. Connection with Helix Theory.

Note that any exceptional collection $(A_0, A_1, \ldots, A_k)$ can be extended to an exceptional collection $\sigma = \{A_i\}$, $i \in \mathbf{Z}$, by requiring σ to be periodic in the sense that

$$A_{i-k-1} = L_{A_{i-k}} L_{A_{i-k+1}} \cdots L_{A_{i-1}} A_i = L^{(k)} A_i.$$

Such a collection σ is uniquely determined by any of its foundations, i.e. by a subcollection of the form $(A_i, A_{i+1}, \ldots, A_{i+k})$ and all the foundations σ generate the same admissible subcategory. Helices can be mutated just like exceptional collections, but the mutations should be performed simultaneously at all the pairs differing by the period. To any helix we can associate a functor

$$\varkappa_T : T \longrightarrow T$$

induced by the change of basis of T given by the rule $A_i \rightarrow L^{(k)} A_i$. One can show (see the paper of Bondal in this collection) that this functor gives Serre duality in T.

In particular, we can consider the case when the helix is full, i.e. generates the whole category $D(X)$. Then $\varkappa$ must be the functor of twisting by the canonical sheaf and so we arrive at the standard definition of a helix—the same as in [2].

Finally, note that the fundamental problem of Helix Theory is the problem of constructibility, in our approach this becomes the question of how many normal bases exist in an admissible category modulo the action of the *elementary transformations*: mutations of pairs. Similarly, the question of whether an exceptional collection ε can be inserted into a large exceptional collection is the question of extending a normal basis to a larger one, or, in other words, the question of the admissibility of the category $L_T(D(X))$ for a given admissible category T.

References

[1] BEILINSON, A.A., BERNSTEIN, I.N. & DELIGNE, P., Faiseaux Pervers, *Asterisque*, **100** (1981)

[2] GORODENTSEV, A.L. & RUDAKOV, A.N., Exceptional Vector Bundles on Projective Space, *Duke Math. J.*, **54** (1987) 115–130.

[3] KAPRANOV, M.M., On the Derived Categories of Coherent Sheaves on some Homogeneous Spaces, *Inv. Math*, **92** (1988) 479–508.

[4] MUKAI, S., On the Moduli Spaces of Bundles on K3 Surfaces, I, in *Vector Bundles* ed. Atiyah et al, Oxford Univ. Press, Bombay, (1986) 341–413.

[5] VERDIER, J.L., Categories Derivée, in *SGA $4\frac{1}{2}$* ed. P. Deligne, Springer-Verlag LNM 569, , (1977) 262–311.

8.   Helices, Representations of Quivers and Koszul Algebras

A. I. Bondal

1.   Introduction.

The aim of this paper is to study functors from various categories to categories of representations of finite-dimensional associative algebras. These functors arise in the following way.

Let $\mathcal{E}$ be an exceptional object of some abelian category $\mathcal{A}$. This means that $\mathrm{Ext}^i(\mathcal{E}, \mathcal{E}) = 0$ for $i > 0$. Then, using $\mathcal{E}$, one can construct a functor $F_{\mathcal{E}}$ from the category $\mathcal{A}$ to the derived category $D^b(\mathbf{mod}\text{-}A)$ of the category of representations of the category $A = \mathrm{Hom}(\mathcal{E}, \mathcal{E})$:

$$F_{\mathcal{E}}(\mathcal{M}) = \mathbf{R}\mathrm{Hom}(\mathcal{E}, \mathcal{M}).$$

$F_{\mathcal{E}}(\mathcal{M})$ is a complex of right A-modules. The functor $F_{\mathcal{E}}$ can be extended to a derived functor $D^b F$ from $D^b \mathcal{A}$ to $D^b(\mathbf{mod}\text{-}A)$. If $\mathcal{E}$ has enough projectives, then $D^b F_{\mathcal{E}}$ is an equivalence of triangulated categories.

As an example one can consider $\mathcal{A} = Ch(\mathbf{P}^n)$, the category of coherent sheaves on projective space $\mathbf{P}^n$. In [1] Beilinson proved that if one defines $\mathcal{E}_0 = \bigoplus_{i=0}^{n} \mathcal{O}(i)$, then $D^b F_{\mathcal{E}_0}$ is an equivalence of categories. Following this, Drezet [6], Gorodentsev and Rudakov [8], constructed a whole series of exceptional bundles which are obtained by successively mutating the bundle $\mathcal{E}_0$. Here, it is more convenient to interpret $\mathcal{E}_0$ as a whole collection of exceptional bundles $\mathcal{O}(i)$ and to mutate within this collection.

Another example of an exceptional collection is the set of projective modules over a finite-dimensional associative algebra. The mutations of such a collection generalize reflective functors [3] and tilting modules [5] which are used in the representation theory of quivers. Generally speaking, the property of a collection being exceptional, as defined in [8], is not preserved by mutating. Therefore, it is necessary to weaken it. In this form it can be successfully used in any triangulated category.

If $i : \mathcal{B} \to \mathcal{A}$, is an inclusion of a subcategory, generated by elements of an exceptional collection, to the base category, then, as will be shown in Theorem 3.2, there exists an adjoint functor $i^* : \mathcal{A} \to \mathcal{B}$

$$\mathrm{Hom}_{\mathcal{A}}(iX, Y) = \mathrm{Hom}_{\mathcal{B}}(X, i^* Y),$$

where $X \in \mathcal{B}$, $Y \in \mathcal{A}$.

This functor is a variant of the *Bar* construction in [10]. In [8] it was observed that mutations of an exceptional collection generate a helix. Theorem 4.1 shows that this is connected with the fact that the entire collection is exceptional.

Later on we demonstrate, via examples, the connection between geometry and algebra, that is, between coherent sheaves on varieties and representations of algebras. For example, sheaves on $\mathbf{P}^1$ correspond to representations of a quiver composed of two vertices and two arrows from one vertex to the other. As is well known, it is a tame quiver and its representations were described by Kronecker. The unique parameter on which the irreducible representations depend takes values on the projective line.

Finally, we examine the problem of determining, from a purely algebraic point of view, when mutations of strongly exceptional collections remain strongly exceptional. The restrictions one has to impose are certain homological conditions on the algebra of homomorphisms between elements of an exceptional collection. We call algebras satisfying these conditions *self-consistent*. The independent study of these algebras is obviously of considerable interest.

2.   Mutations.

As was mentioned in the introduction, in order to be able to work in a general situation, one has to introduce a weaker notion of exceptional collection than given in [8] (see paper 1 in this collection).

Let A be some triangulated category and let A and B be objects of $\mathcal{A}$. For convenience we introduce the following notation

$$\mathrm{RHom}(A, B) = \bigoplus_{k \in \mathbf{Z}} \mathrm{Hom}\big(A, B[k]\big)[-k].$$

This sum is regarded as a graded complex of vector spaces with a trivial differential. The numbers in the square brackets are multiple shifts.

DEFINITION. An object E is called *exceptional* if it satisfies the following conditions:

$$\mathrm{Hom}\big(E, E[k]\big) = 0, \qquad \text{for} \quad k \neq 0, \qquad \mathrm{Hom}(E, E) = \mathbf{C}.$$

DEFINITION. An ordered collection $(E_0, \ldots, E_n)$ of exceptional objects of $\mathcal{A}$ is called an *exceptional collection* if it satisfies the condition $\mathrm{RHom}(E_j, E_k) = 0$ for $j > k$. An exceptional collection consisting of two objects is called an *exceptional pair*.

Let (E, F) be an exceptional pair. We define objects $L_E F$ and $R_F E$ using the following distinguished triangles in the category $\mathcal{A}$:

$$L_E F \longrightarrow \mathbf{RHom}(E, F) \otimes E \longrightarrow F$$

$$E \longrightarrow \mathbf{RHom}(E, F)^* \otimes F \longrightarrow R_F E \tag{1}$$

Duals of vector spaces have the reversed grading. The *left* (respectively *right*) *mutation* of the exceptional pair $\tau = (E, F)$ is defined to be the pair $L_F \tau = (LF, E)$ (repectively, $R_E \tau = (F, RE)$). The lower indices will be dropped if this does not lead to confusion. A mutation of an exceptional collection $\sigma = (E_0, \dots, E_n)$ is defined by mutating a pair of neighbouring objects in this collection:

$$R_{E_i} \sigma = (E_0, \dots, E_{i-1}, E_{i+1}, R_{E_{i+1}} E_i, E_{i+2}, \dots, E_n)$$

$$L_{E_{i+1}} \sigma = (E_0, \dots, E_{i-1}, L_{E_i} E_{i+1}, E_i, E_{i+2}, \dots, E_n) \tag{2}$$

Properties of mutations are described in [8]. It is convenient to interpret the object $R_{E_{i+1}} E_i$ as a shift of E_i to the right in the collection $\sigma = (E_0, \dots, E_n)$. In the mutated collection $R_{E_{i+1}} \sigma$ one can again mutate. In particular, one can shift further to the right. The result of a multifold shift of the object E_i in the collection σ will be denoted by $R_\sigma^k E_i$ and the resulting collection by $R_{E_i}^k \sigma$. Similarly for left mutations.

PROPOSITION 2.1. *A mutation of an exceptional collection is an exceptional collection.*

The proof is presented in the paper of Gorodentsev (Paper 7 in this collection). Let $(X_0, \dots, X_n)$ be a collection of objects in $\mathcal{A}$. We denote the minimal full triangulated subcategory containing the objects X_i by $\langle X_0, \dots, X_n \rangle$. We say that the collection of objects $(X_0, \dots, X_n)$ *generates* the category $\mathcal{A}$ if $\langle X_0, \dots, X_n \rangle$ coincides with $\mathcal{A}$.

LEMMA 2.2. *If the exceptional collection $(E_0, \dots, E_n)$ generates the category $\mathcal{A}$, then a mutated collection also generates $\mathcal{A}$.*

PROOF. Let $\mathcal{A} = \langle E_0, \dots, E_n \rangle$ and $\mathcal{B} = \langle E_0, \dots, E_{i+1}, RE_i, \dots, E_n \rangle$.

$$\mathbf{RHom}(E_i, E_{i+1})^* \otimes E_{i+1}$$

belongs to $\mathcal{A}$ because $\mathcal{A}$ is closed with respect to direct sums and shifts. From this and formula (1), $RE_i \in \mathcal{A}$, and so $\mathcal{B} \subset \mathcal{A}$. Similarly one can see that $\mathcal{A} \subset \mathcal{B}$, because the collection $(E_0, \dots, E_n)$ is obtained from $(E_0, \dots, E_{i+1}, RE_i, \dots, E_n)$ by a left mutation of the pair (E_{i+1}, RE_i).

3.   The Adjoint Functor.

DEFINITION (See [16]). Two triangulated subcategories B and C of a triangulated category A are called *orthogonal* if for any objects $X \in B$ and $Y \in C$, $\operatorname{Hom}_A(X,Y) = 0$.

DEFINITION.   Let B be a triangulated subcategory. The full subcategory generated by the objects $Y \in A$ which satisfy the condition that $\operatorname{Hom}_A(X,Y) = 0$ for all $X \in B$ is called the *right orthogonal complement* of B and is denoted by $B^\perp$. Similarly for *left orthogonal complements*.

LEMMA 3.1. *Let B and C be two orthogonal categories in A and suppose that for any object $X \in A$ there is a distinguished triangle $B \to X \to C$, where $B \in B$ and $C \in C$. Then the inclusion functor $i : B \to A$ (respectively, $j : C \to A$) has a right (left) adjoint functor $i^* : A \to B$ ($j^! : A \to C$), which satisfies the condition $\operatorname{Hom}_A(iB, A) = \operatorname{Hom}_B(B, i^*A)$ for any $A \in A$ and $B \in B$ (similarly for $j^!$).*

PROOF. For any object $X \in A$, we need to construct an object $i^*X \in B$. We have the distinguished triangle $B \to X \to C$. We shall show that such a triangle is unique up to a unique isomorphism. Let $B' \to X \to C'$ be another such triangle. Then applying the functor $\operatorname{Hom}(B', *)$ to the first triangle we see that $\operatorname{Hom}(B', B) = \operatorname{Hom}(B', X)$. Applying this equality to the morphism $\phi \in \operatorname{Hom}(B', X)$ of the second triangle, we obtain a unique morphism $\psi : B' \to B$ which fits into the commutative diagram

$$
\begin{array}{ccccc}
B & \longrightarrow & X & \longrightarrow & C \\
\psi \uparrow & & \uparrow \text{id} & & \\
B' & \longrightarrow & X & \longrightarrow & C'
\end{array}
$$

The inverse morphism and canonical isomorphism from C to C' are constructed similarly. Put $i^*X = B$. Applying the functor $\operatorname{Hom}(T, *)$ to the triangle $B \to X \to C$, where T is an arbitrary object of B, we see that $\operatorname{Hom}(T, i^*X) \xrightarrow{\sim} \operatorname{Hom}(T, X)$.

THEOREM 3.2. *Let $B = \langle E_0, \ldots, E_n \rangle$ be a subcategory of A, and $C = B^\perp$. Then the inclusion functor $i : B \to A$ (respectively, $j : C \to A$) has a right (left) adjoint functor $i^* : A \to B$ ($j^! : A \to C$). Moreover, $j^!X = L^{n+1}X[n+1]$, where $L^{n+1}X$ is defined by induction:*

$$
L^0 X = X, \qquad L^{k+1}X \to \operatorname{Hom}(E_{n-k}, L^k X) \otimes E_{n-k} \to L^k X \xrightarrow{\alpha} L^{k+1}X[1]. \qquad (3)
$$

PROOF. From (3) it follows easily by induction that $L^{n+1}X \in \mathcal{C}$. According to Lemma 3.1 one must check that any object $X \in \mathcal{A}$ can be embedded into a triangle

$$B \longrightarrow X \longrightarrow L^{n+1}X[n+1], \qquad \text{where } B \in \mathcal{B}. \tag{4}$$

We shall prove this by induction on the length of the collection. For $k = -1$ the statement is obvious. Assume that the statement is true for the collection $(E_1, \ldots, E_n)$. We shall check it for the collection $(E_0, \ldots, E_n)$. For any $X \in \mathcal{A}$ we have

$$L^n X[n-1] \longrightarrow B_0 \longrightarrow X \xrightarrow{\beta} L^n X[n], \tag{5}$$

where $B_0 \in \langle E_1, \ldots, E_n \rangle$. Let $\gamma = \alpha[n] \circ \beta$, where α is the morphism from triangle (3) and β is from (5). Then we have the following commutative diagram (see [2], page 24):

$$
\begin{array}{ccccc}
\Phi & \longrightarrow & 0 & \longrightarrow & \Phi[1] \\
\uparrow & & \uparrow & & \uparrow \\
B & \longrightarrow & X & \xrightarrow{\gamma} & L^{n+1}X[n+1] \\
\uparrow & & \uparrow{\scriptstyle\text{id}} & & \uparrow{\scriptstyle\alpha[n]} \\
B_0 & \longrightarrow & X & \xrightarrow{\beta} & L^n X[n]
\end{array}
$$

Since B_0 and Φ belong to $\mathcal{B}$, the middle row of the diagram is a triangle of the form shown in (4).

4.   Helices.

An interesting example of the triangulated category $\mathcal{A}$ is the derived category $D^b(Ch(X))$ of the category of coherent sheaves on a variety X. As an example of an exceptional collection on projective space $\mathbf{P}^n$ we have the collection of sheaves $\mathcal{O}(i)$, $i = 0, 1, \ldots, n$. Mutations of this collection were studied in [8].

Let $(E_0, \ldots, E_n)$ be an exceptional collection. We extend it to infinity in both directions to obtain the sequence $(E_i, \ i = -\infty, \ldots, +\infty)$ of objects of $\mathcal{A}$, defined by induction:

$$
\begin{aligned}
E_{n+i} &= R^n E_i \\
E_{-i} &= L^n E_{n-i}, \qquad \text{where } i > 0.
\end{aligned} \tag{6}
$$

It was shown in [8] that exceptional collections of bundles on $\mathbf{P}^n$, constructed by mutating the collection $\{\mathcal{O}(i)\}$, have the following property: $E_i = E_{i+n+1}(K)$, where E_i, $i = -\infty, \ldots, +\infty$, are to be understood in the above sense.

Such an infinite sequence was called a helix in that paper. We shall extend this definition to an arbitrary variety.

DEFINITION. A sequence E_i, infinite in both directions, of objects of the derived category $D^b(Ch(X))$ of coherent sheaves on a variety X of dimension m, will be called a *helix of period* $n + 1$ if

$$E_i = E_{i+n} \otimes K[m - n],$$

where K is the canonical sheaf and the number in square brackets denote the amount that the object is shifted to the left as a graded complex in $D^b(Ch(X))$.

In the case of projective space $\mathbf{P}^m$, the period n is equal to $m + 1$ and shifts in the derived category do not occur. For a quadric Q^m either there is a shift one place to the right if m is odd or there is no shift if m is even.

Using resolutions of the diagonal, Beilinson showed that the collection $\mathcal{O}(i)$ generates the category $D^b(Ch(\mathbf{P}^n))$, and Drezet, Gorodentsev and Rudakov have shown that this is also true for mutated collections.

DEFINITION. An exceptional collection is called a *foundation* of a helix if the sequence constructed by formulae (6) is a helix of period $n + 1$.

R. G. Swan [15] constructed exceptional collections on quadrics, and Kapranov did the same on Grassmanians, quadrics and flag varieties [10]. We now prove that these collections are also foundations of a helix.

THEOREM 4.1. *Let* $(E_0, \ldots, E_n)$ *be an exceptional collection of bundles on a variety* X *of dimension* m *with very ample anticanonical class. Then the following are equivalent:*

(1) the collection E_i *generates the derived category* $D^b(Ch(X))$;

(2) the collection E_i *is a foundation of a helix.*

We first prove

PROPOSITION 4.2. *The functor* $\mathbf{RHom}(E_n, *)^*$ *is representable in the subcategory* $\langle E_0, \ldots, E_n \rangle$, *generated by the objects* E_i *and the object representing it is* $L^n E_n[n]$, *that is*

$$\mathbf{RHom}(E_n, X)^* \xrightarrow{\sim} \mathbf{RHom}(X, L^n E_n[n]),   \qquad (7)$$

where $X \in \langle E_0, \ldots, E_n \rangle$.

PROOF. Consider the subcategory $\mathcal{B} = \langle E_0, \ldots, E_{n-1} \rangle$. According to Theorem 3.2 there is a triangle $B \to E_n \to L^n E_n[n]$, where $B \in \mathcal{B}$, and $L^n E_n[n] \in \mathcal{B}^\perp$. We see from it that $\mathbf{RHom}(E_n, L^n E_n) = \mathbf{RHom}(E_n, E_n) = \mathbf{C}$. The last equality means that $\mathrm{Hom}(E_n, L^n E_n[k]) = 0$ for $k \neq 0$, and $\mathrm{Hom}(E_n, L^n E_n) = \mathbf{C}$. We shall construct a pairing between $\mathrm{Hom}(E_n, X)$ and $\mathrm{Hom}(X, L^n E_n)$ mapping two morphisms to their composite.

$$\mathrm{Hom}(E_n, X) \otimes \mathrm{Hom}(X, L^n E_n[n]) \longrightarrow \mathrm{Hom}(E_n, L^n E_n[n]).   \qquad (8)$$

We show that this pairing is nondegenerate for any X. From Theorem 3.2, there exists a triangle $Y \to X \to L^n X[n]$ for X, where $Y \in \mathcal{B}$. Using this triangle we can rewrite (8) in the form

$$\mathrm{Hom}(E_n, L^n X[n]) \otimes \mathrm{Hom}(L^n X[n], L^n E_n[n]) \longrightarrow \mathrm{Hom}(E_n, L^n E_n[n]).$$

$L^n X[n]$ belongs to $\mathcal{B}^\perp$. If we consider the exceptional collection $(L^n E_n, E_0, \ldots, E_{n-1})$ then it is clear that $\mathcal{B}^\perp = \langle L^n E_n \rangle$ (for more details see Lemma 6.1). The category $\langle L^n E_n \rangle$ is composed of direct sums and shifts of the object $L^n E_n$, since it is exceptional. In this way it is enough to check the nondegeneracy of (8) for $X = L^n E_n$, but for this case it is obvious.

PROOF. (of Theorem 4.1). 1. $\Rightarrow$ 2. Let $(E_0, \ldots, E_n)$ be a collection satisfying the assumptions of the theorem. According to Proposition 4.2, $L^n E_n[n]$ represents the functor $\mathrm{RHom}(E_n, X)^*$ in the category $D^b(Ch(X))$. But Serre duality implies that

$$\mathrm{RHom}(E_n, X)^* = \mathrm{RHom}(X, E_n \otimes K[m]).$$

From the uniqueness of the representing object, it follows that

$$L^n E_n[n] = E_n(K)[m],$$

and from this

$$E_{-1} = L^n E_n = E_n(K)[m - n].$$

This is the helix condition for $i = -1$. Since any successive subcollection of length $n + 1$ in sequence (6) is exceptional, we have the helix property for any i.

2. $\Rightarrow$ 1. Suppose that the collection $(E_0, \ldots, E_n)$ satisfies condition 2 of the theorem. From Lemma 2.2, mutations of the objects E_i belong to a subcategory $\mathcal{A} = \langle E_0, \ldots, E_n \rangle$. From this it follows that all helices belong to $\mathcal{A}$. Taking into account the invariance of $\mathcal{A}$ under shifts, we have $E_0(pK) \in \mathcal{A}$ for all $p \in \mathbf{Z}$.

The existence of an adjoint functor (see §3) implies that any object X of the domain category can be inserted into the distinguished triangle, $i^* X \to X \to Y$, where $i^* X \in \mathcal{A}$, and $\mathrm{RHom}(A, Y) = 0$ for all $A \in \mathcal{A}$. In this way, it suffices to prove that all such Y are equal to zero. Y is represented by some finite complex $C^\bullet$ of objects of an abelian category of coherent sheaves on X. We shall embed X into projective space $\mathbf{P}^l$, using $-K$: $\phi : X \to \mathbf{P}^l$. Then $\phi_*(C^i)$ be coherent sheaves on $\mathbf{P}^l$ and the operation of tensoring with $-K$ becomes the operation of tensoring with $\mathcal{O}_{\mathbf{P}^l}$. We have

$$\mathrm{RHom}(E_0(pK), C_i) = H_X^*(E_0^* \otimes_{\mathcal{O}(X)} C^i(-pK)) = H_{\mathbf{P}^l}^*(\phi_*(E_0^* \otimes_{\mathcal{O}(X)} C^i \otimes \mathcal{O}(p))$$

For $p \gg 0$ all the higher H^i $(i > 0)$ are be zero. This means that $\mathrm{RHom}(E_0(pK), C^\bullet)$ can be computed by using the complex

$$K^i(p) = H^0_{\mathbf{P}^l}\big(\phi_*(E_0^* \otimes_{\mathcal{O}(X)} C^i) \otimes \mathcal{O}(p)\big)$$

with the natural differential. But $\mathrm{RHom}(E_0(pK), C^\bullet) = 0$, so the complex $K^\bullet$ is acyclic.

The Serre Theorem [14], says that the abelian category $Ch(\mathbf{P}^l)$ of sheaves on $\mathbf{P}^l = \mathbf{P}(V^{l+1})$ is isomorphic to the quotient category of finite dimensional graded modules over $S^\bullet(V^*)$ by a full subcategory of finite dimensional modules. Moreover, this isomorphism takes the sheaf $\mathcal{F}$ to the graded module $\bigoplus_p H^0\big(\mathcal{F} \otimes \mathcal{O}(p)\big)$, where we can assume that $p \gg 0$. Isomorphisms of categories induce isomorphisms of derived categories. Since the complex $\bigoplus_{p \gg 0} K^\bullet(p)$ is acyclic, the corresponding object of the derived category $D^b(S^\bullet)$ equals zero and from the Serre Theorem $\phi_*(E_0^* \otimes C^\bullet) = 0$. From this we have $C^\bullet = 0$, which proves the theorem.

5. Quivers.

Let $\mathbf{C}$ be the complex number field. A *quiver* Δ is a set composed of vertices and arrows between them. We shall be interested in finite quivers, that is, quivers in which the number of points and arrows is finite. A *path* is a sequence of arrows in which the end of any arrow joins with the start of the next arrow. The length of a path is the number of arrows in it. Composition of paths is the composite path (if it is defined). Formal linear combinations with coefficients in the field $\mathbf{C}$ form a path algebra $\mathbf{C}\Delta$ with respect to the operation of composition of paths. Here, multiplication of paths α and β is taken to be equal to zero if the start of β does not coincide with the end of α. The vertices correspond to degenerate paths of length 0. They are the projections in the algebra of paths.

If $S \subset \mathbf{C}\Delta$ is some subset, then the quotient algebra of the path algebra $\mathbf{C}\Delta$ by the ideal generated by S will be called a *quiver with relations*. Generators of the ideal can be chosen in the form of linear combinations of paths with the same start and end points. We shall let $\mathbf{C}\Delta^k$ denote the ideal of $\mathbf{C}\Delta$ generated by paths of length greater than or equal to k.

Quivers with relations represent a wide class of algebras.

Let A be a finite dimensional associative algebra over the field $\mathbf{C}$. An algebra A will be called *a basic algebra* if $A/\mathrm{rad}\,A$ is a direct sum of several copies of $\mathbf{C}$, where $\mathrm{rad}\,A$ is the radical of the algebra A. By Morito, every algebra A' is equivalent to some basic algebra A. This means that the categories of their representations are equivalent. There is a theorem (Gabriel) which says that any basic finite-dimensional

C-algebra is a quiver Δ with relations S. Moreover, a quiver is uniquely determined if one assumes $S \subset C\Delta^k$. Later on we shall assume that this condition is satisfied.

EXAMPLE 5.1. The quiver P_n.

Composed of two vertices and n arrows from the first vertex to the second vertex. For example P_2 : $\bullet \rightrightarrows \bullet$

EXAMPLE 5.2. The quiver A_n.

Composed of n vertices $X_1, \ldots, X_n$ and $n-1$ arrows $\phi_i : X_i \to X_{i+1}$, $i = 1, \ldots, n-1$. For example A_3 : $\bullet \longrightarrow \bullet \longrightarrow \bullet$

EXAMPLE 5.3. The quiver S_n.

Composed of n vertices $X_1, \ldots, X_n$ and $(n-1) \cdot n$ arrows $\phi_i^j : X_i \to X_{i+1}$, $i = 1, \ldots, n-1; j = 1, \ldots, n$. The relations are $\phi_{i+1}^j \phi_i^k = \phi_{i+1}^k \phi_i^j$.

For example S_3 :   with relations $\psi_i \phi_j = \psi_j \phi_i$, $i, j = 1, 2, 3$.

Note that A_2 coincides with P_1, and S_2 coincides with P_2.

EXAMPLE 5.4.   with relations $\beta_j \alpha_i = \delta_i \gamma_j$ for $i, j = 1, 2$.

Let A be the algebra of paths of the quiver Δ with relations S, $A = C\Delta/(S)$. Then the image of $C\Delta^1$ under the corresponding epimorphism $C\Delta \to A$ is the radical of A. In addition, this radical contains the paths of length 0. They can be numbered by vertices of the quiver Δ and will be denoted by $p_\alpha \in A$, where α is a vertex of Δ. The elements p_α are the orthogonal projections; $p_\alpha p_\beta = p_\beta p_\alpha = 0$ for $\alpha \neq \beta$, and $p_\alpha^2 = p_\alpha$.

A representation of a quiver is an A-module, that is a vector space V over the field C with a left action of the algebra A. The action of the orthogonal projections decomposes V into a direct sum $V = \bigoplus_\alpha p_\alpha V$. Put $V_\alpha = p_\alpha V$. Then an arrow from a vertex α to a vertex β gives a linear operator $V_\alpha \to V_\beta$. From this it is clear that our definition of a representation of a quiver coincides with the traditional one, where to every vertex α one associates a vector space V_α and to every arrow from α to β a morphism $V_\alpha \to V_\beta$ in such a way that the relations S are satisfied.

Right A-modules correspond to representations of quivers on Δ^{opp}. They are obtained from Δ by reversing the arrows and rewriting the relations in reverse order.

The representation theory of quivers is concerned with finding which quivers with relations have a finite number of representations and with describing these representations. However, quivers with infinite dimensional representations do exist and such that the problem of classifying their representations does not reduce to the problem of classifying pairs of commuting matrices. The former quivers are called *tame*. The latter quivers are called *wild*. Tame algebras form a very small island in the sea of wild algebras (see [11], [12]).

The quiver A_n from example 5.2 has a finite number of representations. According to the theorem of Gabriel [7], finitely representable quivers without relations correspond to Coxeter graphs [4] without multiple edges.

An example of a tame quiver is P_2 from example 5.1.

An *ordered quiver with relations* is, by definition, a quiver whose vertices are ordered and whose arrows go from smaller to larger numbers (except for arrows of length 0). The quivers in examples 5.1, 5.2, 5.3, 5.4 are all ordered. In example 5.4 one can do this by placing the upper and lower vertices in an arbitrary order between the left and right vertices.

Let Δ be an ordered quiver with vertices $X_0, \ldots, X_n$, and let p_i be the projection corresponding to X_i in the algebra A of paths of the quiver Δ. Later on it will be convenient to consider right modules over A. Let A-**mod** denote the category generated by these modules. Each representation V of the algebra A admits a decomposition $V = \bigoplus_{i=0}^{n} G_i V$, where $G_i V = V p_i$. Let S_i denote the representation for which $G_j V = 0$ for $j \neq i$ and $G_i V = C$, and all the arrows are represented by non-zero morphisms. The modules S_i ($i = 0, 1, \ldots, n$) describe all the irreducible representations of A. Indeed, any module V has a filtration by the modules

$$F^k V = \bigoplus_{i=0}^{k} G_i V.$$

The quotient $F^k V / F^{k-1} V$ is a direct sum of several copies of S_k.

Projective modules of the algebra A are submodules of A regarded as right modules over themselves and take the form $P_k = p_k A$. There is a decomposition $A = \bigoplus_{i=0}^{n} P_i$. In addition,

$$A = \mathrm{Hom}_A(A, A) = \mathrm{Hom}_A\left(\bigoplus_{i=0}^{n} P_i, \bigoplus_{i=0}^{n} P_i\right)$$
$$= \bigoplus_{i,j} \mathrm{Hom}(P_i, P_j).$$

This equality allows us to interpret arrows of a quiver as maps between projective modules. In particular, $\mathrm{Hom}(P_i, P_j) = 0$ for $i > j$.

It is easy to check that $G_l P_k = p_k A p_l = 0$ and $G_k P_k$ is a one-dimensional space. This makes it possible to construct an exact sequence

$$0 \longrightarrow F^{k-1} P_k \longrightarrow P_k \longrightarrow S_k \longrightarrow 0. \tag{9}$$

LEMMA 5.5 *Let V be a right A-module. Assume that $G_i V = 0$ for $i > k$. Then $V \in \langle P_0, \ldots, P_k \rangle$.*

Recall that $\langle P_0, \ldots, P_k \rangle$ is the triangulated category generated by the objects $P_0, \ldots, P_k$ in $D^b(\text{mod-}A)$.

PROOF. We use induction. For $k = 0$ the statement is obvious. If it is true for $k = s-1$, then for any V, $F^{s-1} V \in \langle P_0, \ldots, P_{s-1} \rangle$. For V satisfying the assumptions of the lemma with $k = s$, we have an exact sequence $0 \to F^{s-1}V \to V \to G_s V \to 0$, where $G_s V$ is regarded as several copies of S_s. But from (9) it follows that $S_k \in \langle P_0, \ldots, P_s \rangle$. From this $V \in \langle P_0, \ldots, P_s \rangle$.

LEMMA 5.6 *Let $\mathcal{B} = \langle P_0, \ldots, P_{k-1} \rangle$ be a subcategory of $D^b(\text{mod-}A)$, and $i : \mathcal{B} \to \mathcal{A}$ the inclusion functor. Then*

$$i^! P_k = S_k = L^k P_k[k]. \tag{10}$$

PROOF. Consider the exact sequence in (9). In this sequence, $F^{k-1} P_k$ belongs to $\mathcal{B}$ by Lemma 5.5. This sequence can be viewed as a triangle in $D^b(\text{mod-}A)$. According to Lemma 3.1, it suffices to convince ourselves that $S^k \in \mathcal{B}^\perp$. Since P_i are projective modules we have $\text{Ext}^i(P_i, S_k) = 0$ for $j \neq 0$. It remains to prove that $\text{Hom}(P_i, S_k) = 0$ for $i < k$.

Any such homomorphism gives rise to a collection of graded components of the map: $G_j P_i \to G_j S_k$. But the only nonzero component of S_k is the kth and $G_k P_i = 0$ for $i < k$. This means that all the morphisms are zero. By Theorem 3.2 we get the equations in (10).

6. Functors in the Category $D^b(\text{mod-}A)$ related to an Exceptional Collection.

As before let $\mathcal{A}$ be a trangulated derived category.

DEFINITION. The collection $(E_0, \ldots, E_n)$ of objects of $\mathcal{A}$ which satisfy

a) $\text{Hom}(E_i, E_j[k]) = 0$,   for all i, j; $k \neq 0$, $k \in \mathbf{Z}$,

b) $\text{Hom}(E_i, E_j) = 0$,   for all $i > j$,

will be called a *strongly exceptional collection*.

An example of a strongly exceptional collection is $(\mathcal{O}, \mathcal{O}(1), \ldots, \mathcal{O}(n))$ on $\mathbf{P}^n$.

LEMMA 6.1. *Suppose that $\mathcal{A}$ is generated by the (not necessarily strongly) exceptional collection $(E_0, \ldots, E_n)$. Set $\mathcal{C} = \langle E_0, \ldots, E_k \rangle$ and $\mathcal{B} = \langle E_{k+1}, \ldots, E_n \rangle$. Then $\mathcal{C} = \mathcal{B}^\perp$.*

PROOF. Evidently, C is orthogonal to B on the right. Let $X \in B^\perp$ and let $i : C \to A$ be the inclusion functor. We have a triangle $i^*X \to X \to Z$ with $Z \in C^\perp$. As i^*X and X belong to $B^\perp$, $Z \in B^\perp$. Therefore Z is orthogonal to all the generators E_i and so equals 0.

Let $E = \bigoplus_{i=0}^{n} E_i$ and $A = \operatorname{Hom}(E, E)$. A is the path algebra of a finite ordered quiver with relations. This quiver has $n + 1$ vertices, and for projective modules P_i over this algebra we have isomorphisms $\operatorname{Hom}_A(E_i, E_j) \overset{\sim}{\to} \operatorname{Hom}_A(P_i, P_j)$.

THEOREM 6.2. *The category $A = \langle E_0, \ldots, E_n \rangle$, generated by the elements of a strongly exceptional collection, is equivalent to the derived category $D^b(\mathbf{mod}\text{-}A)$ of right modules over the algebra A.*

PROOF. The functor Φ, giving the equivalence of the categories, is constructed by induction. In the kth step, we assume the existence of an equivalence Φ_k between $\langle E_{n-k}, \ldots, E_n \rangle$ and $\langle P_{n-k}, \ldots, P_n \rangle$, and an isomorphism

$$\phi_k^i : \operatorname{Hom}(E_i, X) \overset{\sim}{\to} \operatorname{Hom}(P_i, \Phi_k X),$$

where $X \in \langle E_{n-k}, \ldots, E_n \rangle$, compatible with Φ_k.

For $k = -1$ the theorem is clear. Suppose that Φ_{k-1} and ϕ_{k-1} have already been constructed. From Theorem 3.2 for arbitrary $X \in \langle E_{n-k}, \ldots, E_n \rangle$ we have a distinguished triangle

$$Y \longrightarrow X \longrightarrow Z \overset{d}{\longrightarrow} Y[1], \tag{11}$$

where $Y \in \langle E_{n-k+1}, \ldots, E_n \rangle$ and $Z \in \langle E_{n-k+1}, \ldots, E_n \rangle^\perp$.

From Lemma 6.1, $Z \in \langle E_{n-k} \rangle$. The category $\langle E_{n-k} \rangle$ consists of direct sums and shifts of the object E_{n-k}. Put $\Phi_k(E_{n-k}) = P_{n-k}$, then it is easy to extend Φ_k to the subcategory $\langle E_{n-k} \rangle$.

Define Φ_k from the triangle

$$\Phi_k X \longrightarrow \Phi_k Z \overset{\phi_{k-1}(d)}{\longrightarrow} \Phi_{k-1} Y[1].$$

From the unique triangle of the form in (11), it follows that $\Phi_k X$ is well defined. Now let $\phi : X \to X'$ be a morphism of objects of the category $\langle E_{n-k}, \ldots, E_n \rangle$. Consider their associated triangles of the form in (11), i.e. $Y \to X \to Z$ and $Y' \to X' \to Z'$. Applying the functor $\mathbf{RHom}$ to these triangles we obtain a commuting diagram

$$
\begin{array}{ccccc}
\operatorname{RHom}(Z, Z') & \overset{d}{\longrightarrow} & \operatorname{RHom}(X, Z') & \longrightarrow & \operatorname{RHom}(Y, Z') \\
\uparrow & & \uparrow & & \uparrow \\
\operatorname{RHom}(Z, X') & \longrightarrow & \operatorname{RHom}(X, X') & \longrightarrow & \operatorname{RHom}(Y, X') \\
\uparrow & & \uparrow & & \uparrow \\
\operatorname{RHom}(Z, Y') & \longrightarrow & \operatorname{RHom}(X, Y') & \longrightarrow & \operatorname{RHom}(Y, Y')
\end{array}
$$

Here, the columns and rows are distinguished triangles in the category of complexes of vector spaces. Φ_k is defined on $\mathbf{RHom}(Y,Y')$ since Y and $Y' \in \langle E_{n-k+1}, \ldots, E_n \rangle$. It is defined on $\mathbf{RHom}(Z,Y')$ using ϕ_k, since $Z \in \langle E_{n-k} \rangle$ and $Y \in \langle E_{n-k+1}, \ldots, E_n \rangle$. From the bottom horizontal triangle we see that it extends uniquely to $\mathbf{RHom}(X,Y')$. But we have $\mathbf{RHom}(Y,Z') = 0$, which means that $\alpha : \mathbf{RHom}(Z,Z') \to \mathbf{RHom}(X,Z')$ is an isomorphism. Consequently, Φ_k is defined on $\mathbf{RHom}(X,Z')$ and hence is uniquely determined on maps from X to X'. Applying the functor $\mathbf{RHom}(E_i, *)$ to the triangle in (11), we see that the morphisms ϕ_k^i are uniquely determined on $\mathrm{Hom}(E_i, X)$. It is easy to convince oneself that Φ_k is an equivalence of the corresponding subcategories, and that ϕ_k^i are compatible with Φ_k.

If we assume that $\mathcal{A}$ is a derived category of some abelian category, then the functor $\mathbf{RHom}(E, *)$ is an equivalence of categories. Note also that it is possible to reformulate the statement of the theorem in terms of an arbitrary exceptional collection, but we shall not need this.

We shall show how to realize this equivalence of categories in the case of coherent sheaves on varieties.

EXAMPLE 6.3. Let $\mathcal{A} = D^b(Ch(\mathbf{P}^1))$. As a strongly exceptional collection we take $(\mathcal{O}, \mathcal{O}(1))$. $A = \mathrm{Hom}(\mathcal{O} \oplus \mathcal{O}(1), \mathcal{O} \oplus \mathcal{O}(1))$. A is the algebra of paths of the quiver P_2 in example 5.1. This quiver is composed of two vertices and two arrows from the first vertex to the second. The projections p_1 and p_2 correspond to identity operators on $\mathcal{O}$ and $\mathcal{O}(1)$. Irreducible (right) modules of the algebra A have been described by Kronecker and are well known. They correspond to the roots in the weight lattice (λ_1, λ_2), here, λ_1 is a multiple of an irreducible representation of S_1 and λ_2 is a multiple of the representation S_2 in the Jordan-Hölder composition series. The real roots have weights of the form $(n, n+1)$ and $(n+1, n)$, where $n \geq 0$. In accordance with Theorem 6.2, the complexes $\mathcal{O}(n)$ and $\mathcal{O}(-n)[1]$, of coherent sheaves on $\mathbf{P}^1$ correspond respectively to these roots. Imaginary roots are the weights (n, n), $n > 0$. To each such root there corresponds a one-dimensional family of irreducible representations, $V_{n,x}$, where $x \in \mathbf{P}^1$. We assign to the module $V_{n,x}$ the sheaf of jets at x up to the $(n-1)^{\mathrm{st}}$ order inclusively.

For any quiver without relations (this corresponds to the homological dimension of A being equal to 1), one can show that all the irreducible objects of the category $D^b(\mathbf{mod}\text{-}A)$ up to a shift, are equivalent to pure modules (that is, complexes of the form $0 \to M \to 0$).

EXAMPLE 6.4. Consider the exceptional collection $(\mathcal{O}, \ldots, \mathcal{O}(k))$ in $D^b(Ch(\mathbf{P}^k))$. The algebra corresponding to this collection is

$$A = \bigoplus_{i,j} \mathrm{Hom}(\mathcal{O}(i), \mathcal{O}(j)),$$

which is the path algebra of the quiver in example 5.3.

EXAMPLE 6.5. Let Q be a nondegenerate quadric in $\mathbf{P}^3$. It is known that Q is isomorphic to $\mathbf{P}^1 \times \mathbf{P}^1$. As the exceptional collection one can take $(\mathcal{O}, \mathcal{O}(0,1), \mathcal{O}(1,0), \mathcal{O}(1,1))$. Here, $\mathcal{O}(i,j) = \mathcal{O}(i) \boxtimes \mathcal{O}(j)$. Mutations of this collection were studied in [13]. This collection corresponds to the algebra from example 5.4. Exceptional collections on quadrics of arbitrary dimension can be found in [15] and [9].

7. Koszul Algebras.

Let $(E_0, \ldots, E_n)$ be a strongly exceptional collection. We consider the algebra $A = \bigoplus_{i,j=0}^{n} \operatorname{Hom}(E_i, E_j)$. This is also an algebra of paths of an ordered quiver of $n+1$ vertices with relations. Let A_k denote the subspace of A generated by paths of length k. Then A is a graded algebra

$$A = A_0 \oplus A_1 \oplus A_2 \oplus \cdots$$

A_0 is a subalgebra which consists of one-dimensional algebras:

$$A_0 = \bigoplus_{i=0}^{n} \mathbf{C} p_i,$$

where p_i is the ith orthogonal projection. The A_i are equipped with an A_0-bimodule structure. We shall assume that the arrows of a quiver exist only between neighbouring vertices. This is equivalent to the fact that A is generated by A_0 and A_1. In other words we have the following equalities:

$$A_k = A_1 \otimes_{A_0} A_1 \otimes \cdots \otimes A_1 / I_k,$$

where I_k is a space of relations with an A_0-bimodule structure.

The algebra A of the form described above is called *quadratic* if all the relations I_k are generated by $I_2 \subset A_1 \otimes_{A_0} A_1$. The generators in I_2 are linear combinations of paths of length 2 with the same start and end points. Formally, the condition of being quadratic can be expressed in the form of the following inclusion:

$$I_k \subset I_2 \otimes_{A_0} A_1 \otimes \cdots \otimes A_1 + A_1 \otimes I_2 \otimes A_1 \otimes \cdots \otimes A_1 + \cdots + A_1 \otimes \cdots A_1 \otimes I_2,$$

where every term contains $k-1$ factors and every factor can be viewed as a subspace of $A_1^{\otimes k}$.

The dual algebra B of the algebra A can be defined as follows. B is the quadratic algebra in which $B_0 = A_0$, $B_1 = A_1^*$ and $J_2 = I_2^{\perp}$, where $J_2 \subset B_1 \otimes_{A_0} B_1 = A_1^* \otimes_{A_0} A_1^*$, and the orthogonal complement is given by the natural pairing.

The space $K = B^* \otimes_{A_0} A$ comes equipped with the structure of a complex. For an arbitrary A_0-bimodule V we put $V^{ij} = p_i V p_j$. It is evident that $A_1^{i,j} \neq 0$ only for $i = j + 1$, since the elements of A_1 are arrows of a quiver. Consider an arbitrary basis e_i^j of the spaces $A_1^{j+1,j}$ and let ξ_i^j be the dual basis of the spaces $B_1^{j,j+1}$. Consider also the operator $d : K \rightarrow K$, where $d = \sum_{i,j} l(\xi_i^j)^* \otimes l(e_i^j)$, and here, $l(e_i^j)$ is the operation of left multiplication by e_i^j in A. It is easy to check that $d^2 = 0$.

DEFINITION.   The complex $K = B^* \otimes A$, equipped with the differential d is called a *Koszul complex*.

The differential preserves the A_0-bimodule structure of the complex K, and therefore $K^{i,j}$ is invariant with respect to d. The spaces $K^{i,i}$ are one-dimensional for all i and the differential d restricts to zero on them.

DEFINITION.   The algebra A is called *Koszul* if the homology of the complex K is isomorphic to $\bigoplus_{i=0}^{n} K^{i,i}$.

The differential preserves the right A-module structure. This, together with the action of A_0 on the left, makes it possible to decompose K into a sum of complexes of A-modules $K = \bigoplus K^i$, where

$$K^i = \bigoplus_j K^{i,j} = \bigoplus_{k,j} B^{*i,k} \otimes A^{k,j} = \bigoplus_{k,j} B^{*j,k} \otimes P_k,$$

where P_k are projective A-modules, given by $P_k = p_k A$. K is a graded complex of A-modules, where the kth part of the grading is

$$\cdots \longrightarrow B^{*i,i-2} \otimes_{A_0} \otimes P_{i-2} \longrightarrow B^{*i,i-1} \otimes_{A_0} P_{i-1} \xrightarrow{d} P_i \longrightarrow 0.$$

The Koszul condition means that the complexes K^i are exact at all the terms apart from P_i, where the homology is one-dimensional and isomorphic to $P_i^{i,i}$. This means that we have an exact sequence

$$\cdots \rightarrow B^{*i,i-2} \otimes_{A_0} P_{i-2} \rightarrow B^{*i,i-1} \otimes_{A_0} P_{i-1} \rightarrow P_i \rightarrow S_i \rightarrow 0 \qquad (12)$$

where S_i is an irreducible module corresponding to the ith vertex.

PROPOSITION 7.1   *The algebra A is Koszul if and only if $\mathrm{Ext}^k(S_i, S_j) = 0$ for all $k \neq i - j$.*

PROOF. a) Suppose that the algebra is Koszul. In Lemma 5.6 it is proved that $\mathrm{Hom}(P_k, S_j) = 0$ for $k < j$. According to Lemma 5.5 $S_k \in \langle P_0, \ldots, P_k \rangle$, and therefore $\mathrm{Hom}(P_k, S_j) = 0$ for $k > j$. In order to calculate the Ext groups we apply the functor $\mathrm{Hom}(*, S_j)$ to the projective resolution of the module S_i, given by sequence (12). We see that $\mathrm{Ext}^k(S_i, S_j) = 0$ for $k \neq i - j$, and $\mathrm{Ext}^{i-j}(S_i, S_j) = B^{i,j}$.

b) Suppose, for example, that the ith complex of the form (12) is not exact. Let the first place on the right where this complex is not exact be $B^{*i,j} \otimes P_j$. Consider the corresponding piece of complex (12):

$$\cdots \to B^{*i,j-2} \otimes P_{j-2} \to B^{*i,j-1} \otimes P_{j-1} \to \cdots \tag{13}$$

This complex decomposes into a sum of complexes $K^{i,s}$ which are graded components with respect to the right action of A_0:

$$\cdots \to B^{*i,j-1} \otimes P_{j-1}^{j,s} \to B^{*i,j} \otimes P_j^{j,s} \to \cdots$$

For $s = j, j - 1$ this complex is exact at P_j. This follows from the fact that the algebra B is quadratic. Choose the least s with the property that $K^{i,s}$ is not exact at the place we are interested in. Then $s < j - 1$. Let $H_j^{i,s}$ be the homology of this complex at the jth place. Then we can extend the complex (13) to an exact complex of projective modules. Moreover, in the jth and $(j-1)^{\mathrm{st}}$ place the resolution takes the form

$$\cdots \to P \oplus B^{*i,j-1} \otimes P_{j-1} \oplus H_j^{i,s} \otimes P_s \to B^{*i,j} \otimes P_j \to \cdots \to S_i$$

where P is the sum of the projective modules P_t for $t < s < j - 1$. Computing $\mathrm{Ext}^{i-j+1}(S_i, S_s)$ using this resolution, we see that they are equal to $\left(H_j^{i,s}\right)^*$. Since $s < j - 1$ the theorem follows.

It was shown earlier (in Lemma 5.6) that $S_k = L^k P_k[k]$. Thus

$$\mathrm{Hom}^k(L^i P_i, L^j P_j) = \mathrm{Hom}^{k+i-j}(L^i P_i[i], L^j P_j[j]) = \mathrm{Ext}^{k+i-j}(S_i, S_j).$$

Using Theorem 6.2, we obtain the following:

COROLLARY 7.2 *The fact that the algebra of homomorphisms of the strongly exceptional collection* $\{E_i\}$ *is Koszul is equivalent to the fact that the collection* $\{L^i E_i\}$ *is strongly exceptional.*

DEFINITION. The collection $\{L^n E_n, L^{n-1} E_{n-1}, \ldots, E_0\}$ is called the *left dual collection* of the collection $\{E_0, \ldots, E_n\}$. Similarly, we define the *right dual collection* to be $\{E_n, R E_{n-1}, \ldots, R^n E_0\}$.

8. Self-consistent Algebras and Mutations of Strongly Exceptional Collections.

It would be interesting to discover under what the conditions mutations preserve the condition of strongly exceptionality of collections. In the geometric situation (when $\mathcal{A} = D^b(Ch(\mathbf{P}^n))$) the question of whether a collection can be mutated, was studied in [8].

Consider the infinite sequence (6), constructed using the strongly exceptional collection $\sigma = (E_0, \ldots, E_n)$. We shall denote the helix by S_σ. A mutation of a collection will be called *admissible* if the resulting collection is a strongly exceptional collection. A helix is called admissible if the result of a mutation is admissible.

LEMMA 8.1. *The following conditions are equivalent to the condition that a helix S is admissible:*

a) *all the mutations of S of the form R_E^k are admissible, where E is an element of the helix;*

b) *all the mutations of S of the form L_E^k are admissible;*

c) *in any foundation of the helix, mutations inside the foundation of the form R_E^k and L_E^k are admissible mutations of the collection, where E is an element of the foundation.*

When mutating helices it follows from considerations of periodicity that it is enough to restrict ourselves to shifts over distances shorter than the period of the helix.

PROOF. The equivalence of a) and b) follows from periodicity. It is also obvious that a) and b) imply c).

Now let R_E^k be a mutation of the helix S and τ be some foundation containing $R^k E$. If E does not belong to τ then $R^n E$ belongs to τ, and the mutation $R^k E$ can be thought of as $L_{R^n E}^{n-k}$, that is, as a left mutation within the foundation. This proves that a) follows from b).

Claim. If one of the equivalent conditions of the lemma is satisfied, then all of the mutations of the helix are admissible.

It is clear that the conditions of 8.1 are necessary.

We shall now reformulate c) for an exceptional collection in terms of algebras of homomorphisms of this collection.

In analogy with the Koszul complex $K = B^* \otimes A$ (§8), one can define another three complexes associated to the algebra A

$$K_1 = A \otimes B^*, \qquad K_2 = A \otimes B \quad \text{and} \quad K_3 = B \otimes A.$$

The differentials in these complexes are given by the formulae (in the notation of §7):

$$d_1 = \sum r(e_i^j) \otimes r(\xi_i^j)^*$$
$$d_2 = \sum r(e_i^j) \otimes l(\xi_i^j)$$
$$d_3 = \sum r(\xi_i^j) \otimes l(e_i^j).$$

The complex K_1 can be interpreted as a Koszul complex for the algebra A^{opp} and the complex K_2 as the complex K_3 for the algebra A^{opp}.

Consider the complex $K_2 = A \otimes B$. It is bigraded, since it is an A_0-bimodule: $K_2 = K_2^{i,j}$, $K_2^{i,j} = p_i K_2 p_j$. $K_2^{i,j}$ are the graded complexes with respect to the differential d_2

$$0 \to A^{i,j} \to A^{i,j-1} \otimes B^{j-1,j} \to \cdots \to A^{i,k} \otimes B^{k,j} \to \cdots \to A^{i,0} \otimes B^{0,j}$$

DEFINITION. An algebra A is called *co-Koszul* if the complexes $K_2^{i,j}$ for $i \neq j$, are exact at all places other that the last ($A^{i,0} \otimes B^{0,j}$).

Note that in general, these conditions do not extend to the diagonal complexes $K_2^{i,i}$.

DEFINITION. An algebra A is called *self-consistent* if A and A^{opp} are Koszul and co-Koszul.

This means that the complexes $K^{i,j}$ and $K_1^{i,j}$ are exact, for $i \neq j$, and $K_2^{i,j}$ (respectively, $K_3^{i,i}$), are exact at the end (respectively the start) for $i \neq j$.

PROPOSITION 8.2. *The following are equivalent:*

a) *the algebra A of homomorphisms of an exceptional collection σ is self-consistent;*

b) *mutations of the form R_E^k and L_E^k inside σ are admissible.*

PROOF. Following Theorem 6.2, we can identify E_i and P_i. Assume that the algebra is Koszul, then, according to §7

$$B^{j,i} = \mathrm{RHom}(L^i P_i, L^j P_j)$$
$$= \mathrm{RHom}(L^{i-j} P_i, P_j)$$
$$= \mathrm{RHom}(P_i, R^{i-j} P_j)$$
$$= \mathrm{RHom}(R^{i-j-1} P_j, P_i).$$

In particular, as a complex, $\mathrm{RHom}(R^{i-j-1} P_j, P_i)$ is concentrated in degree zero. It follows easily from this that the complexes $K_2^i = \bigoplus_j K_2^{i,j}$ represent the objects

$R^{n-i}P_i$, just as the complexes K^i in §7 represented $L^i P_i = S_i[-i]$. Admissibility of mutations means that there are no Hom^i for $i \neq 0$ between P_j and $R^{n-i}P_i$, where $i \neq j$. If we then compute with the complexes K_2^i and K^i, we can see immediately that the conditions of the lemma are equivalent. However, we have assumed that it is Koszul. Consequently, it remains to show that b) implies that it is Koszul. To do this from the construction in 7.3, we need to convince ourselves that $\text{Hom}^k(L^i P_i, L^j P_j) = 0$ for $k \neq 0$. We have $\textbf{R}\text{Hom}(L^i P_i, L^j P_j) = \textbf{R}\text{Hom}(L^{i-j}P_i, P_j)$, and this implies what is wanted because of b).

The algebras of the quivers in examples 5.1 and 5.3 are self-consistent. The quiver A_n for $n > 2$ provides us with an example of a Koszul complex which is not self-consistent.

We can see from this that the complexes $K_2^{i,i}$ are an interesting characteristic of an algebra. For the collection $\{\mathcal{O}(i)\}$ on $\textbf{P}^n$, they are exact everywhere except at the end. In the general case, the place where they are not exact (if it is the only one) should be regarded as a quantum supergrading.

Now we turn to the geometric case. let $\mathcal{A} = D^b(Ch\, X)$, where X is a variety of dimension n. Suppose that the collection $\sigma = (E_0, \ldots, E_n)$ generates $\mathcal{A}$.

THEOREM 8.3. *If the collection σ and its left dual consists of bundles and are strongly exceptional collections, then any mutation of σ is composed of bundles and the resulting collection is strongly exceptional.*

PROOF. It is necessary to convince ourselves of the fact that, in a helix S_σ, the results of mutations of the form L_E^k are bundles. Then, using the methods of [8] for the collection $\{\mathcal{O}(i)\}$ on $\textbf{P}^n$, we can prove that mutations of collections remain strongly exceptional.

Let $A = \bigoplus_{i,j} \text{Hom}(E_i, E_j)$. From corollary 7.3 it follows that this algebra is Koszul. Theorem 6.2 shows that we can pass from $D^b(Ch\, X)$ to $D^b(\textbf{mod-}A)$, and from the collection $\{E_i\}$ to the collection $\{P_i\}$ of projective right A-modules. Then $L^n P_n = S_n[-n]$, because of formula (12), can be represented by the complex K^n. Now we turn to $D^b(Ch\, X)$. To apply this to the complex K^n, we need to replace P_i with E_i

$$0 \to B^{*n,0} \otimes E_0 \to \cdots \to B^{*n,n-1} \otimes E_{n-1} \to E_n \to 0.$$

This complex represents $L^n E_n[n]$. But since the collection $\{E_i\}$ generates the category, Theorem 4.1 implies that we have $L^n E_n[n] = E_n \otimes K$, where K is the canonical class. That is, the complex is exact at all places except for the first. The partial mutations $L^k E_n$ are components of this complex and, consequently, are also represented by bundles. We can prove by induction, in the same way as in [8], that

$L_{E_n}^k \sigma$ is a strongly exceptional collection . This means that the foundation on the left, neighbouring the foundation σ in the helix S_σ, is a strongly exceptional collection. An elementary mutation of a collection corresponds to a right mutation of the dual collection. Therefore a dual collection is also strongly exceptional. Using corollary 7.3, we see that the foundation to the left of σ in the helix S_σ satisfies the conditions of the theorem. We can also convince ourselves that the bundle furthest to the right (that is, E_{n-1}) can be shifted. Iterating this process, we can deduce the existence of left mutations in the helix for all the bundles to the left of E_n. Similarly, all this can be done for right mutations, since a right dual collection is a left dual collection twisted by the anticanonical class.

This theorem can be applied not only to the collection $\{\mathcal{O}(i)\}$ on $\mathbf{P}^n$, but also to exceptional collections on odd-dimensional quadrics [13], [9]. In order to be able to apply this theorem to even-dimensional quadrics and Grassmanians, on which there are also exceptional collections [10], it is necessary to modify the notion of an exceptional object in a collection. Namely, an exceptional object in this new sense should be the direct sum of several objects which are exceptional in the former sense of the word and which have no RHom's between them (for all pairs). The rest of the theory can be developed in an identical way. Collections for quadrics and Grassmanians, as was shown in [10] satisfy the assumptions of Theorem 8.3.

Finally, I would like to thank A. Gorodentsev, M. Kapranov, I. Panin for numerous discussions and A.I. Kostrikin, A.N. Rudakov, A.N. Tyurin for their interest and support.

References

[1] BEILINSON, A.A., Coherent Sheaves on $\mathbf{P}^n$ and Problems in Linear Algebra, *Funk. An.*, **12** (1978) 68–69.

[2] BEILINSON, A.A., BERNSTEIN, I.N. & DELIGNE, P., Faiseaux Pervers, *Asterisque*, **100** (1981)

[3] BERNSTEIN, I.N., GELFAND, I.M. & PONOMARYOV, V.A., Coxeter Functors and the Gabriel Theorem, *Russian Math. Surveys* , **28** (1973) 17–32.

[4] BOURBAKI, N., *Lie Groups and Lie Algebras*, Chapters 4–6.

[5] BRENNER, S., & BUTLER, M.K.R., Generalization of Beilinson-Gabriel-Ponomaryov Reflection Functors, in *Representation Theory II, Ottowa* ed. V. Dlab, P. Gabriel, Springer-Verlag LNM 832, (1980).

[6] DREZET, J-M., Fibrés Exceptionelle et Suite Spectrale de Beilinson Generalise Sur $\mathbf{P}_2(\mathbf{C})$, *Math. Ann.*, **B275** (1986) 25–48.

[7] GABRIEL, P., Unzerleghare Derstellungen I, *Manuscripta Math.*, **6** (1972) 71–103.

[8] GORODENTSEV, A.L. & RUDAKOV, A.N., Exceptional Vector Bundles on Projective Space, *Duke Math. J.*, **54** (1987) 115–130.

[9] KAPRANOV, M.M., On the Derived Category of Coherent Sheaves on Grassmanian Manifolds, *Math. USSR Isv.*, **24** (1985) 183–192.

[10] KAPRANOV, M.M., On the Derived Categories of Coherent Sheaves on some Homogeneous Spaces, *Inv. Math*, **92** (1988) 479–508.

[11] NAZAROVA, A.A. & ROYTER, A.V., Categorical Matrix Problems and Brouwer-Trol Problems, in *Research in Representation Theory*, Notes Sc. Seminar LOMY **28** (1972).

[12] RINGEL, C.M., *Tame Algebras and Integral Quadratic Forms*, LNM 1099, Springer-Verlag (1984).

[13] RUDAKOV, A.N., Exceptional Bundles on a Quadric, *Math. USSR Isv.*, **33** (1989) 115–138

[14] SERRE, J-P., Coherent Algebraic Sheaves, in *Fibred Spaces and their Application* (1958) 372–458.

[15] SWAN, R.G., K-Theory of Quadric Hypersurfaces, *An. Math.*, **122** (1985) 113–153.

[16] VERDIER, J.L., Categories Derivée, in *SGA $4\frac{1}{2}$* ed. P. Deligne, Springer-Verlag LNM 569, (1977) 262–311.

9. Exceptional Collections on Ruled Surfaces

A.B. Kvichansky   D.Yu. Nogin

In the study of exceptional bundles on surfaces, the model examples are $\mathbf{P}^2$ and $\mathbf{P}^1 \times \mathbf{P}^1$. These surfaces are homogeneous varieties on which every exceptional sheaf is locally free. In this paper we will consider exceptional coherent sheaves on rational ruled surfaces H_e which may not be locally free. As was established by Gorodentsev and Rudakov [2] [3] [4], all the exceptional bundles on $\mathbf{P}^2$ and $\mathbf{P}^1 \times \mathbf{P}^1$ are obtained by mutations of helices composed of line bundles.

We will be interested in whether there exist helices with nice properties, composed of line bundles, on ruled surfaces. For reasons which we shall not dwell on here, we shall only be interested in helices of period four (since $\operatorname{rk} K_0(H_e) = 4$). To study such helices we consider exceptional collections of length four.

1. The Geometry of the Base.

As is well known, a rational ruled surface has the structure $\mathbf{P}(E)$ for some rank two bundle E over $\mathbf{P}^1$. Every such surface is characterised by a non-negative integer e—we shall denote $\mathbf{P}(\mathcal{O}_{\mathbf{P}^1} \oplus \mathcal{O}_{\mathbf{P}^1}(-e))$ by H_e. The group $\operatorname{Pic} H_e$ has two generators: F, which is the fibre of the natural projection, and C, a specially chosen section. These satisfy the relations

$$C^2 = -e, \qquad C \cdot F = 1 \quad \text{and} \quad F^2 = 0.$$

The divisor $aC + bF$ will be denoted by (a, b) and the corresponding line bundle by $\mathcal{O}(a, b)$.

The divisor (a, b) is ample if and only if $a > 0$ and $b > ae$. The anti-canonical class $-K_{H_e}$ is equal to $(2, 2 + e)$ and is ample only for $e = 0, 1$.

2. Cohomology of Invertible Sheaves.

On any surface S, knowing $h^0(L)$ for all $L \in \operatorname{Pic} S$ makes it possible to calculate the rest of the cohomology. Indeed, $h^2(L) = h^0(K_S - L)$ and $h^1(L) = h^0(L) + h^2(L) - \chi(L)$, where, when S is rational, $\chi(L) = \frac{1}{2}L(L - K_S) + 1$. In our case, we calculate $h^0(L)$ as follows:

LEMMA 2.1. *On the ruled surface H_e, with $e \geqslant 1$, we have $h^0(\mathcal{O}(a, b)) = 0$, for $a < 0$ and $b < 0$, and, in the remaining cases, $h^0(\mathcal{O}(a, b)) = \chi(\mathcal{O}(\tilde{a}, b))$, where $\tilde{a} = a$, if $a \leqslant b/e$, and $\tilde{a} = [b/e]$, if $a > b/e$.*

PROOF. The first part of the lemma is obvious. For $0 < a < b/e$, the sheaf $\mathcal{O}(a,b)$ is ample and all its cohomology is concentrated in H^0, i.e. $h^0(\mathcal{O}(a,b)) = \chi(\mathcal{O}(a,b))$.

In the case $a > b/e > 0$, we consider the sequence

$$0 \to \mathcal{O}(-1,0) \to \mathcal{O} \to \mathcal{O}_C \to 0,$$

associated to the curve $C = (1,0)$, and we tensor it by $\mathcal{O}(a,b)$. Here $\mathcal{O}_C \otimes \mathcal{O}(a,b) = \mathcal{O}_C(b - ae)$ and $b - ae < 0$. But then $h^0(\mathcal{O}_C(b - ae)) = 0$ and from the sequence

$$0 \to \mathcal{O}(a-1,b) \to \mathcal{O}(a,b) \to \mathcal{O}_C(b-ae) \to 0$$

we get $h^0(\mathcal{O}(a,b)) = h^0(\mathcal{O}(a-1,b))$. One can extend this to get a chain of equalities

$$h^0(\mathcal{O}(a,b)) = h^0(\mathcal{O}(a-1,b)) = \cdots = h^0(\mathcal{O}(a-k,b))$$

for all k for which $a - k + 1 > b/e$. Thus, $h^0(\mathcal{O}(a,b)) = h^0(\mathcal{O}(\tilde{a},b))$, where $\tilde{a} = \min\{a \mid a + 1 > b/e\} = [b/e]$. If we also have $\tilde{a} < b/e$, then $\mathcal{O}(\tilde{a},b)$ is ample and $h^0(\mathcal{O}(a,b)) = \chi(\mathcal{O}(\tilde{a},b))$. If however $\tilde{a} = b/e$, then $\mathcal{O}(\tilde{a}-1,b)$ is ample and the equality $h^0(\tilde{a},b) = \chi(\tilde{a},b)$ follows from the fact that, in the short exact sequence

$$0 \to \mathcal{O}(\tilde{a}-1,b) \to \mathcal{O}(\tilde{a},b) \to \mathcal{O}_C(b-\tilde{a}e) \to 0,$$

the first and last terms have nonzero cohomology.

The remaining case $a = 0$ is obvious.

3.   Exceptional Sheaves and Exceptional Bundles.

Since the ruled surface H_e has the structure of $\mathbf{P}(E)$ for a rank two bundle E on $\mathbf{P}^1$, all the automorphisms of $\mathbf{P}^1$ and the automorphisms of E determine the group of automorphisms of the surface H_e. The set of points of H_e splits into two orbits under the action of this group, namely $C = (1,0)$ and $H_e \setminus C$. It is obvious from this that every rigid sheaf F on H_e is either locally free or has singularity set C, and moreover

$$S_0(F) = \{x \in H_e \mid \mathrm{hd} F_x = 2\} = \varnothing.$$

This implies that, if a rigid sheaf is not locally free, then its torsion sheaf $T(F)$ is nonzero, $T(F)$ is a bundle over C and $F/T(F)$ has no singularities, so is equal to F^{**}.

In particular, every exceptional sheaf of rank zero is an exceptional sheaf on $C = \mathbf{P}^1$, i.e. it is equal to $\mathcal{O}_C(d)$ for some integer d. From the structure sequence

$$0 \to \mathcal{O}(-1,0) \to \mathcal{O} \to \mathcal{O}_C \to 0$$

we deduce that, for $e \geqslant 1$, the sheaf $\mathcal{O}_C$ is simple, rigid and has $^2\langle \mathcal{O}_C \mid \mathcal{O}_C \rangle = e - 1$.

4. Regular Exceptional Collections.

An *exceptional pair* is an ordered pair (E, F) of exceptional sheaves for which $^i\langle F \mid E \rangle = 0$ for all i. An exceptional pair is *regular* if at most one of the spaces $^0\langle E \mid F \rangle$ and $^1\langle E \mid F \rangle$ is nonzero. As previously shown, every exceptional pair on $\mathbf{P}^2$ and $\mathbf{P}^1 \times \mathbf{P}^1$ is regular—see Gorodentsev [1]. On H_e for $e \geqslant 2$ there exist irregular exceptional pairs, e.g. $(\mathcal{O}, \mathcal{O}(1, 0))$.

Since we can only mutate regular exceptional pairs, we shall be interested in helices, every foundation of which is a regular collection. We call a collection regular if every ordered pair in it is regular.

We shall look for regular collections of invertible sheaves $\{E_i\}$ on H_e for $e \geqslant 1$. One can specify such collections by giving the initial term E_0 and the successive increments $L_i = E_i \otimes E_{i-1}^*$. Without loss of generality we may assume that $E_0 = \mathcal{O}$. The increments should satisfy the conditions that $H^j(L_i^*) = {}^j\langle E_i \mid E_{i-1} \rangle = 0$ for all j, and that at most one of the spaces $H^0(L_i)$ and $H^1(L_i)$ is nonzero.

The condition $H^j(L_i^*) = 0$ gives $\chi(L_i^*) = \frac{1}{2}L_i(L_i + K_{H_e}) + 1 = 0$. If $L_i = \mathcal{O}(a, b)$ we then obtain

$$\chi(L_i^*) = (a - 1)\left(b - 1 - \frac{1}{2}ae\right) = 0,$$

i.e. $a = 1$ or $b = \frac{1}{2}ae + 1$. For $a \geqslant 3$ we have $b \geqslant \frac{3}{2}e + 1$ and then

$$h^2\big(\mathcal{O}(-a, -b)\big) = h^0\big(\mathcal{O}(a - 2, b - 2 - e)\big) \neq 0.$$

Indeed, for $e \geqslant 2$ we have $b - 2 - e \geqslant \frac{1}{2}e - 1 \geqslant 0$ and for $e = 1$ we get $b \geqslant \frac{3}{2} + 1$, i.e. $b \geqslant 3$ and $b - 2 - e \geqslant 0$. Similarly, for $a \leqslant -1$ we get $h^0\big(\mathcal{O}(-a, -b)\big) \neq 0$.

Thus the only sheaves $L = \mathcal{O}(a, b)$ for which the pair $(\mathcal{O}, L)$ is exceptional are $\mathcal{O}(0, 1)$, $\mathcal{O}(1, b)$ for any b and $\mathcal{O}(2, e + 1)$. We can now ask whether the pair $(\mathcal{O}, L)$ is regular. For $L = \mathcal{O}(0, 1)$ and for $L = \mathcal{O}(1, b)$, with $b < 0$ or $b \geqslant e$, it clearly is. We now calculate h^0 and h^1 for the remaining cases.

For $\mathcal{O}(1, b)$ with $0 \leqslant b \leqslant e - 1$, i.e. $[b/e] = 0$, we have

$$h^0\big(\mathcal{O}(1, b)\big) = \chi\big(\mathcal{O}(0, b)\big) = b + 1 \quad \text{and} \quad h^1\big(\mathcal{O}(1, b)\big) = e - b - 1.$$

Thus, for $b = e - 1$ the pair $(\mathcal{O}, \mathcal{O}(1, b))$ is regular, while for $0 \leqslant b \leqslant e - 2$ it is not. The calculation for the sheaf $L = \mathcal{O}(2, e + 1)$ shows that for $e = 1, 2$ we have $h^0(L) = 6$ and $h^1(L) = 0$, while for $e > 2$ we have $h^0(L) = e + 4$ and $h^1(L) = e - 2$. Thus, finally, we arrive at the following:

PROPOSITION 4.1.

a) *If* $e = 1$, *all exceptional pairs* $(\mathcal{O}, L)$ *are regular, where* L *is* $\mathcal{O}(0, 1)$, $\mathcal{O}(1, b)$ *for* $b \in \mathbf{Z}$ *or* $\mathcal{O}(2, 2)$.

b) *If* $e = 2$, *the only irregular exceptional pair is the one with* $L = \mathcal{O}(1, 0)$. *Hence the increment in a regular pair is* $\mathcal{O}(0, 1)$, $\mathcal{O}(1, b)$ *for* $b \in \mathbf{Z} \setminus \{0\}$ *or* $\mathcal{O}(2, 3)$.

c) *If* $e \geqslant 3$, *regular pairs have increments* $\mathcal{O}(0, 1)$ *or* $\mathcal{O}(1, b)$ *for* $b \in \mathbf{Z} \setminus \{0, \dots, e-2\}$. *The pairs* $\mathcal{O}(1, b)$ *for* $0 \leqslant b \leqslant e - 2$ *and* $\mathcal{O}(2, e+1)$ *are exceptional but irregular.*

It remains for us to construct regular collections from regular pairs. Examining the various possibilities yields the following regular collections of length four:

On H_1,

type N_k:	$\mathcal{O}$, $\mathcal{O}(0, 1)$, $\mathcal{O}(1, k)$, $\mathcal{O}(1, k+1)$	for any k,
type Y_k:	$\mathcal{O}$, $\mathcal{O}(1, k)$, $\mathcal{O}(1, k+1)$, $\mathcal{O}(2, 2)$	for any k.

On H_2,

type N_k:	$\mathcal{O}$, $\mathcal{O}(0, 1)$, $\mathcal{O}(1, k)$, $\mathcal{O}(1, k+1)$	for $k \geqslant 2$ or $k \leqslant -2$,
type Y_k:	$\mathcal{O}$, $\mathcal{O}(1, k)$, $\mathcal{O}(1, k+1)$, $\mathcal{O}(2, 3)$	for $k = 1$, $k \geqslant 4$ or $k \leqslant -2$.

On H_e for $e \geqslant 3$,

type N_k:	$\mathcal{O}$, $\mathcal{O}(0, 1)$, $\mathcal{O}(1, k)$, $\mathcal{O}(1, k+1)$	for $k \geqslant e$ or $k \leqslant -2$.

5.   Helices on Ruled Surfaces.

We shall now examine which regular collections (E_0, E_1, E_2, E_3) on H_e determine periodic collections $\{E_i\}_{i \in \mathbf{Z}}$, satisfying $E_{i-4} = E_i(K_{H_e})$, which are helices. To start with, we need every foundation to be a regular collection. This condition is not satified for any regular collection on H_e with $e \geqslant 3$, nor for the regular collections on H_2 of type N_k, with $k = 3, 4, 5$, and Y_k, with $k = -2, 4$. In particular, we have the following:

PROPOSITION 5.1. *On the surfaces* H_e *with* $e \geqslant 3$, *there are no helices of period four composed of invertible sheaves.*

Note that the periodic collections obtained by extending the collections N_k and Y_k differ only by a renumbering of the terms, so it is sufficient to just consider helices arising from collections of type Y_k. Now, in the collection Y_k, the left shift of $\mathcal{O}(1, k+1)$ is $\mathcal{O}(1, k-1)$ and the right shift of $\mathcal{O}(1, k)$ is $\mathcal{O}(1, k+2)$, and so, on the surface H_2, the shifts $L^{(2)}\mathcal{O}(1, 2)$ and $R^{(2)}\mathcal{O}(1, 1)$ are not defined for the collection Y_1.

We shall now show that extending the collection Y_k gives a helix on H_1 for any k, and on H_2 for $k \geqslant 5$ or $k \leqslant -3$.

LEMMA 5.2. *On the surface* H_e, *with* $e = 0, 1$ *or* 2, *the right shift* $R^{(2)}\mathcal{O}$ *is well-defined for the collection* $(\mathcal{O}, \mathcal{O}(1, k), \mathcal{O}(1, k+1))$ *if* $k \geqslant e + 1$ *or* $k \leqslant -2$. *Indeed,* $R^{(2)}\mathcal{O} = \mathcal{O}(2, e)$.

PROOF. If $k \geqslant e + 1$, then the divisor $(1, k)$ is ample and its linear system defines an inclusion of H_e into the projective space $\mathbf{P}^N$, with $N = h^0(\mathcal{O}(1, k)) - 1 = 2k + 1 - e$. Restricting the Euler sequence

$$0 \to \mathcal{O}_{\mathbf{P}^N} \to (N + 1)\mathcal{O}_{\mathbf{P}^N}(1) \to T\mathbf{P}^N \to 0$$

to the image of the inclusion, we obtain the exact sequence

$$0 \to \mathcal{O}_{H_e} \to H^0(\mathcal{O}(1, k))^* \otimes \mathcal{O}_{H_e}(1, k) \to E \to 0,$$

in which the first morphism is canonical, i.e. $E = R^{(1)}\mathcal{O}$. Note that $E = T\mathbf{P}^N|_{H_e}$ is a bundle with $\mathrm{rk}(E) = 2k + 1 - e$ and $c_1(E) = (2k + 2 - e, (2k + 2 - e)k)$. Now consider the regular pair $(\mathcal{O}(1, k + 1), \mathcal{O}(2, e))$ and the universal extension

$$0 \to \mathcal{O}(2, e) \to \tilde{E} \to (2k - e)\mathcal{O}(1, k + 1) \to 0,$$

where $2k - e = {}^1\langle \mathcal{O}(1, k+1) \mid \mathcal{O}(2, e)\rangle$. From this sequence, we find that the topological invariants of $\tilde{E}$ are $\mathrm{rk}(\tilde{E}) = 2k + 1 - e$ and $c_1(\tilde{E}) = (2k + 2 - e, (2k + 2 - e)k)$.

For $e = 0, 1, 2$ the divisor $-K_{H_e}$ is effective and $h^0(-K_{H_e}) = 9 \geqslant 2$. As was shown by Gorodentsev [1], these conditions are sufficient for all exceptional bundles to be stable with respect to the polarisation given by $-K_{H_e}$. From general properties of stable bundles we deduce that $E = \tilde{E}$, i.e. the last sequence defines a shift of E through $\mathcal{O}(1, k + 1)$ of type "recoil".

If $k \leqslant -2$, we can consider the dual collection twisted by $\mathcal{O}(2, e)$. But then we are in the case just considered, because $-(k + 1) + e \geqslant e + 1$.

PROPOSITION 5.3. *The periodic collection over H_2 obtained by extending the collection Y_k is a helix when $k \geqslant 5$ or $k \leqslant -3$.*

PROOF. We have already seen that, in such a collection, every foundation is regular. It remains to observe that for any sheaf E from such a collection τ, the mutation $R_E^{(3)}\tau$ is the composition of a shift of $\mathcal{O}(a', b')$ through $\mathcal{O}(a', b' + 1)$, giving an increment $\mathcal{O}(0, 2)$, and a shift of $\mathcal{O}(a, b)$ through the pair $(\mathcal{O}(a + 1, b + k), \mathcal{O}(a + 1, b + k + 1))$ for $k \leqslant -3$ or $k \geqslant 4$, which, by Lemma 5.2, gives an increment $\mathcal{O}(2, e)$. Consequently, we get a total increment $\mathcal{O}(2, e + 2) = \mathcal{O}(-K_{H_e})$, as required.

REMARK 5.4. On the surface H_2, not every mutation of a helix is a helix. For example, from the collection Y_{-3}, after three right mutations of the pair $(\mathcal{O}(1, -3), \mathcal{O}(1, -2))$, we get a collection in which the right shift of $\mathcal{O}$ is not defined. This is because the triangle equations are not satisfied on H_2 (see paper 1 in this collection). This difficulty might be overcome if we were able to expand the list of possible types of mutations, to include a method of mutating irregular pairs.

PROPOSITION 5.5. *The periodic collections over H_1 obtained by extending the collections Y_k are helices.*

PROOF. All such collections are periodic and regular. To prove the proposition, it is sufficient to demonstrate that for sheaves on H_1 the condition of Lemma 5.2 holds for all $k \in \mathbf{Z}$. The rest of the proof is then a direct copy of the proof of Proposition 5.3. Lemma 5.2 says that for $|k| \geqslant 2$ we can define the right shift $R^{(2)}\mathcal{O}$ for the collection $(\mathcal{O}, \mathcal{O}(1,k), \mathcal{O}(1,k+1))$ over H_1, and that it is equal to $\mathcal{O}(2,1)$. It remains to check this for $k = -1, 0, 1$. Note that, as in Lemma 5.2, the case $k = -1$ can be obtained from the case $k = 1$ by considering the twisted dual collection.

For $k = 0$, we have the collection $(\mathcal{O}, \mathcal{O}(1,0), \mathcal{O}(1,1))$. Here, ${}^0\langle \mathcal{O} \mid \mathcal{O}(1,0) \rangle = 1$ and the right shift $R^{(1)}\mathcal{O}$ is defined by the structure sequence

$$0 \to \mathcal{O} \to \mathcal{O}(1,0) \to \mathcal{O}_C(-1) \to 0.$$

It is easy to check that $(\mathcal{O}_C(-1), \mathcal{O}(1,1))$ is a regular pair and that ${}^1\langle \mathcal{O}_C \mid \mathcal{O}(1,1) \rangle = 1$. This means that the structure sequence

$$0 \to \mathcal{O}(1,1) \to \mathcal{O}(2,1) \to \mathcal{O}_C(-1) \to 0$$

determines a universal extension, i.e. $R^{(2)}\mathcal{O} = R^{(1)}\mathcal{O}_C(-1) = \mathcal{O}(2,1)$, as required.

In the case $k = 1$ we use a different argument. We think of H_1 as $\mathbf{P}^2$ with a point blown up and we consider the natural projection $\pi : H_1 \to \mathbf{P}^2$. For arbitrary bundles E and F on $\mathbf{P}^2$ we have the projection formula:

$$
\begin{aligned}
\mathrm{R}^i \pi_* \big(\mathrm{Hom}(\pi^* E, \pi^* F) \big) &= \mathrm{R}^i \pi_* (\pi^* E^* \otimes \pi^* F) \\
&= \mathrm{R}^i \pi_* (\pi^* E^*) \otimes F \\
&= \mathrm{R}^i \pi_* (\mathcal{O}_{H_1}) \otimes \mathrm{Hom}(E, F) = 0 \quad \text{for } i \geqslant 1
\end{aligned}
$$

and $\pi_* \big(\mathrm{Hom}(\pi^* E, \pi^* F) \big) = \mathrm{Hom}(E, F)$. Then, from the Leray spectral sequence, we deduce that

$$H^i \big(\mathrm{Hom}(\pi^* E, \pi^* F) \big) = H^i \big(\mathrm{Hom}(E, F) \big), \text{ i.e. } {}^i \langle \pi^* E \mid \pi^* F \rangle = {}^i \langle E \mid F \rangle.$$

It is clear from this that lifting from $\mathbf{P}^2$ to H_1 takes exceptional bundles to exceptional bundles and that exceptional pairs are also preserved. Furthermore, since there is a natural isomorphism $\mathrm{Hom}(\pi^* E, \pi^* F) \cong \mathrm{Hom}(E, F)$, all mutations of bundles on $\mathbf{P}^2$ lift to H_1 and so helices also lift.

In particular, the Euler sequence

$$0 \to \mathcal{O}_{\mathbf{P}^2} \to 3\mathcal{O}_{\mathbf{P}^2}(1) \to T\mathbf{P}^2 \to 0$$

becomes the sequence

$$0 \to \mathcal{O}_{H_1} \to 3\mathcal{O}_{H_1}(1,1) \to \pi^* T\mathbf{P}^2 \to 0,$$

which defines the "division" of $\mathcal{O}$ by $\mathcal{O}(1,1)$. Here $\mathrm{rk}(\pi^*T\mathbf{P}^2) = 2$ and $c_1(\pi^*T\mathbf{P}^2) = (3,3)$, which is why $\pi^*T\mathbf{P}^2$ coincides with the bundle E obtained by the universal extension

$$0 \to \mathcal{O}(2,1) \to E \to \mathcal{O}(1,2) \to 0$$

(here $\dim{}^1\langle\mathcal{O}(1,2) \mid \mathcal{O}(2,1)\rangle{=}1$).

In the process of proving this proposition, we have also obtained the following result:

PROPOSITION 5.6. *A helix on* $\mathbf{P}^2$ *lifts to a helix on* H_1.

Thus the set of ranks of exceptional sheaves on H_1 includes all non-negative even numbers (e.g. $\mathcal{O}_C$, $\pi^*T\mathbf{P}^2$, $T\mathbf{P}^{2k}|_{H_1}$ for $k \geqslant 2$), and also all the Markov numbers (c.f. [2] [3]). In order to obtain new exceptional bundles on H_1 by mutations of helices we need to check whether the conditions of the triangle axiom (see p2) are satisfied. This question still remains open.

References

[1] GORODENTSEV, A.L., Exceptional Bundles on Surfaces with a Moving Anti-canonical Class, *Math. USSR Isv.*, **33** (1989) 67–83.

[2] GORODENTSEV, A.L., & RUDAKOV, A.N., Exceptional Vector Bundles on Projective Space, *Duke Math. J.*, **54** (1987) 115–130.

[3] RUDAKOV, A.N., The Markov Numbers and Exceptional Bundles on $\mathbf{P}^2$, *Math. USSR Isv.*, **32** (1989) 99–112.

[4] RUDAKOV, A.N., Exceptional Bundles on a Quadric, *Math. USSR Isv.*, **33** (1989) 115–138.

10. Exceptional Bundles on K3 Surfaces

S. A. Kuleshov

In this paper exceptional bundles on a smooth K3 surface are studied. The case of a K3 surface differs from the cases considered earlier ($\mathbf{P}^2$, $\mathbf{P}^1 \times \mathbf{P}^1$) by having a zero canonical class. Nevertheless, K3 surfaces are regular and have the same property $h^1(\mathcal{O}(\pm D)) = 0$ for any effective divisor D, for which $h^0(\mathcal{O}_D) = 1$. Thanks to the zero canonical class, Serre duality on a K3 surface takes the form $^i\langle E \mid F \rangle^* \cong {}^{2-i}\langle F \mid E \rangle$ for any coherent sheaves.

1. Exceptional Bundles and the Mukai Lattice.

We call a bundle E on a K3 surface *exceptional* if $h^0(E^* \otimes E) = 1$ and $h^1(E^* \otimes E) = 0$. It follows immediately from Serre duality that $h^2(E^* \otimes E) = 1$.

In the study of bundles on a K3 surface it is very convenient to represent coherent sheaves by elements of a lattice, introduced by Mukai in [2].

For a K3 surface X we will introduce a scalar product $(\ |\ )$ on $H^0(X, \mathbf{Z}) \oplus H^2(X, \mathbf{Z}) \oplus H^4(X, \mathbf{Z})$ given by $(u|u') = \alpha \cup \alpha' - r \cup s' - r' \cup s$, for every $u = (r, \alpha, s)$ and $u' = (r', \alpha', s')$, where the symbol "$\cup$" denotes the product operation in the integral cohomology ring. Following Mukai we will denote the lattice we obtain by the symbol $\tilde{H}(X, \mathbf{Z})$. To every coherent sheaf on X we will assign an element of the Mukai lattice:

$$v(E) = \left(r(E),\ c_1(E),\ r(E) - c_2(E) + \frac{c_1(E)}{2} \right).$$

This lattice has the very remarkable property: for any coherent sheaves E and E' on X the following relations are satisfied:

$$v^2(E) = -\chi(E, E); \qquad (v(E)|v(E')) = -\chi(E, E'). \tag{1}$$

Therefore exceptional bundles on a K3 surface give Mukai vectors with square -2.

In his paper, Mukai obtained the following result ([2], Theorem 5.1). Let $v = (r, \ell, s)$ be a primitive isotropic vector in $\tilde{H}^{1,1}(S, \mathbf{Z})$ of rank $r \geqslant 1$, and let A be any ample divisor. Then there exists a simple sheaf E, μ-semistable with respect to A, whose Mukai vector coincides with v. (Here, by a *primitive vector*, we mean a vector which is not divisible by an integer in the lattice.)

One might hope to prove the following conjecture.

Let S be a K3 surface with $\mathrm{Pic}(S) = \mathbf{Z}$, and let v be a primitive vector in the Mukai lattice with square -2. Then there exists an exceptional bundle E such that $v(E) = v$.

2.   Exceptional Pairs and Canonical Maps.

DEFINITION.   An ordered pair of exceptional bundles (A, B) on a K3 surface is called *exceptional* if $^1\langle A \mid B \rangle = 0$ and $^2\langle A \mid B \rangle = 0$. From Serre duality we see that $^1\langle B \mid A \rangle = \,^0\langle B \mid A \rangle = 0$, for an exceptional pair (A, B).

For exceptional pairs in the case of a K3 surface, just as for other surfaces considered in these papers, the following assertion holds:

LEMMA 1.   *Let X be a smooth complex algebraic K3 surface, and let (A, B) be an exceptional pair of bundles on X. Then*

(1) if the canonical map $^0\langle A \mid B \rangle \otimes A \xrightarrow{\alpha} B$ is an epimorphism, then the kernel of α is an exceptional bundle;

(2) if the canonical map $A \xrightarrow{\beta} \,^0\langle A \mid B \rangle^* \otimes B$ is a monomorphism, then the cokernel of β is an exceptional bundle.

PROOF.   We apply the functor $\langle B \mid$ to the exact sequence

$$0 \longrightarrow L \longrightarrow \,^0\langle A \mid B \rangle \otimes A \longrightarrow B \longrightarrow 0 \qquad\qquad (*)$$

to obtain

$$0 \to \,^0\langle B \mid L \rangle \to \,^0\langle A \mid B \rangle \otimes^0\langle B \mid A \rangle \to \,^0\langle B \mid B \rangle \to$$
$$\to \,^1\langle B \mid L \rangle \to^0\langle A \mid B \rangle \otimes \,^1\langle B \mid A \rangle \to \qquad\qquad (**)$$
$$\to \,^1\langle B \mid B \rangle \to \,^2\langle B \mid L \rangle \to^0\langle A \mid B \rangle \otimes \,^2\langle B \mid A \rangle \to \,^2\langle B \mid B \rangle \to 0.$$

Since the pair (A, B) is exceptional, the sequence $(**)$ degenerates into the following:

$$0 \to \,^0\langle B \mid L \rangle \to 0 \to \mathbf{C} \to \,^1\langle B \mid L \rangle \to 0$$
$$0 \to \,^1\langle B \mid L \rangle \to \,^0\langle A \mid B \rangle \otimes \,^2\langle B \mid A \rangle \to \mathbf{C} \to 0$$

from which we get the relations $^0\langle B \mid L \rangle = 0$, $\dim^1\langle B \mid L \rangle = 1$, $\dim^2\langle B \mid L \rangle = \left(\dim^0\langle A \mid B \rangle\right)^2 - 1$. Applying the functor $\langle A \mid$ to $(*)$, we obtain

$$0 \to \,^0\langle A \mid L \rangle \to \,^0\langle A \mid B \rangle \otimes \mathbf{C} \xrightarrow{\sim} \,^0\langle A \mid B \rangle \to \,^1\langle A \mid L \rangle \to 0$$
$$0 \to \,^1\langle A \mid L \rangle \to \,^0\langle A \mid B \rangle \to 0.$$

From these sequences we get the following equalities:

$$^0\langle A \mid L \rangle = \,^1\langle A \mid L \rangle = 0; \qquad \dim^2\langle A \mid L \rangle = \dim^0\langle A \mid B \rangle.$$

We now apply the functor $|L\rangle$ to $(*)$, taking into account the relations we have obtained:

$$0 \to {}^0\langle L \mid L\rangle \to C \to 0$$

$$0 \to {}^1\langle L \mid L\rangle \to {}^2\langle B \mid L\rangle \to {}^0\langle A \mid B\rangle^* \otimes {}^2\langle A \mid L\rangle \to {}^2\langle L \mid L\rangle \to 0.$$

Since Serre duality implies ${}^0\langle L \mid L\rangle \cong {}^2\langle L \mid L\rangle^* \cong C$, we get the equation $d - n^2 + 1 + n^2 - 1 = 0$, where $d = \dim {}^1\langle L \mid L\rangle$ and $n = \dim {}^0\langle A \mid B\rangle$. The second part of the lemma can be proved in a similar way.

Here, it is appropriate to note that for any exceptional bundle E on a smooth K3 surface, there always exists an integer k such that $(\mathcal{O}, E(k))$ is an exceptional pair and it is also possible to left mutate it. Indeed, since $E(k)$ is simple, it follows that either $h^0(E(k))$ or $h^2(E(k)) = 0$. But it is known that one can always choose a number k such that the $E(k)$ is generated by its global sections, and $h^1(E(k)) = 0$.

In order to check whether the canonical maps of exceptional bundles are epimorphic or monomorphic, it is convenient to use the following

LEMMA 2. *Let ... A_0, A_1, A_2, ... be a sequence of exceptional bundles on some variety. Let us assume that for any integer k and $n = 1$ the sequences*

$${}^0\langle A_k \mid A_{k+n}\rangle \otimes A_k \longrightarrow A_{k+n} \longrightarrow 0$$

$$0 \longrightarrow A_k \longrightarrow {}^0\langle A_k \mid A_{k+n}\rangle^* \otimes A_{k+n} \qquad \text{are exact.}$$

Then these sequences are exact for any n.

PROOF. Induct on n. Let us assume that the canonical maps of the pairs (A_k, A_{k+m}) are epimorphic for all $m < n$. Consider the following commutative diagram

$$
\begin{array}{ccc}
& & 0 \\
& & \uparrow \\
{}^0\langle A_k \mid A_{k+n}\rangle \otimes A_k & \xrightarrow{\;\;\text{can}\;\;} & A_{k+n} \\
\uparrow{\scriptstyle \phi \otimes \mathrm{id}} & & \uparrow{\scriptstyle \text{can}} \\
{}^0\langle A_k \mid A_{k+n-1}\rangle \otimes \langle A_{k+n-1} \mid A_{k+n}\rangle \otimes A_k & \longrightarrow & \langle A_{k+n-1} \mid A_{k+n}\rangle \otimes A_{k+n-1} \to 0
\end{array}
$$

where $\phi : {}^0\langle A_k \mid A_{k+n-1}\rangle \otimes \langle A_{k+n-1} \mid A_{k+n}\rangle \longrightarrow {}^0\langle A_k \mid A_{k+n}\rangle$ is the composition map. Since the bottom row and the right column in this diagram are exact, the sequence ${}^0\langle A_k \mid A_{k+n}\rangle \otimes A_k \xrightarrow{\text{can}} A_{k+n} \longrightarrow 0$ is also exact. The canonical map of the pair (A_k, A_{k+n}) is monomorphic if and only if the canonical map of the pair (A_{k+n}^*, A_k^*) is epimorphic and this can be checked by using part of the lemma which we have already proved.

3. Exceptional Bundles on the Dual Plane.

Let $\pi : S \to \mathbf{P}^2$ be a two-fold covering, branched over a smooth curve of degree six. We will assume that the Picard group of the surface S is isomorphic to $\mathbf{Z}$. This condition will hold for a general branch curve. It is not difficult to see that the generator of $\mathrm{Pic}(\mathbf{P}^2)$ lifted to S turns out to be the generator of the Picard group of the dual plane, that is, if ℓ is a line on $\mathbf{P}^2$ then $\pi^*(\ell)$ is a generator of $\mathrm{Pic}\, S$, which we will denote $\mathcal{O}_S(1)$, or simply $\mathcal{O}(1)$, if it is clear which surface we are dealing with. It is known that the Chern classes of bundles on $\mathbf{P}^2$ and their images under a lift to a degree 2 cover are related by the formulae:

$$c_2(\pi^*E) = 2c_2(E)$$
$$c_1(\pi^*E)c_1(\pi^*E') = 2c_1(E)c_1(E'). \tag{2}$$

To construct exceptional bundles on a dual plane, we need the following lemma:

LEMMA 3. *Let A and B be bundles on $\mathbf{P}^2$. Then the following inequality is satisfied*

$$\dim{}^0\langle A \mid B \rangle \leqslant \dim{}^0\langle \pi^*A \mid \pi^*B \rangle \leqslant \dim{}^0\langle A \mid B \rangle + \dim{}^0\langle \pi^*A \mid \pi^*B(-3) \rangle.$$

In particular, if $\dim{}^0\langle \pi^*A \mid \pi^*B(-3) \rangle = 0$, then ${}^0\langle A \mid B \rangle \cong {}^0\langle \pi^*A \mid \pi^*B \rangle$.

PROOF. Set $A' = \pi^*A$, $B' = \pi^*B$. On S there is an involution i which switches branches. Since A' and B' are invariant with respect to this involution, the involution also acts on the space ${}^0\langle A' \mid B' \rangle$ and it determines a direct sum decomposition ${}^0\langle A' \mid B' \rangle = {}^0\langle A' \mid B' \rangle_+ \oplus {}^0\langle A' \mid B' \rangle_-$ with eigenvalues ± 1. It is clear that ${}^0\langle A' \mid B' \rangle_+ \cong {}^0\langle A \mid B \rangle$. Let us consider an arbitrary morphism $\varepsilon \in {}^0\langle A' \mid B' \rangle_-$. Since $i^*\varepsilon = -\varepsilon$, for any pair of points p_1 and p_2 which are identified under π, $\varepsilon|_{A'(p_1)} = -\varepsilon|_{A'(p_2)}$, where $\varepsilon|_{A'(p_i)}$ denotes restriction to a fibre. Therefore ε is zero on the branching divisor. Consequently there is an inclusion ${}^0\langle A' \mid B' \rangle_- \subset {}^0\langle A' \mid B'(-3) \rangle$. This proves the lemma.

REMARK If E and E' are bundles on $\mathbf{P}^2$ then $\chi(E, E') + \chi(E', E) = \chi(\pi^*E, \pi^*E')$. From the Riemann-Roch Theorem, the Euler characteristic of bundles on $\mathbf{P}^2$ can be calculated from the formula

$$\chi(E, E') = \sum_i (-1)^i \dim{}^i\langle E \mid E' \rangle = rr'(p(\mu' - \mu) - \Delta - \Delta'),$$

where $\Delta(E) = \dfrac{1}{r}\left(c_2 - \dfrac{c_1^2}{2} + \dfrac{c_1^2}{2r}\right)$ is the discriminant of E, $\mu(E) = \dfrac{c_1}{r}$ is the slope of E and $p(x) = \dfrac{x^2}{2} + \dfrac{3}{2}x + 1$ [3]. It follows from this formula that

$$\chi(E, E') + \chi(E', E) = 2rr' - 2c_1c_1' - 2c_2r' - 2c_2'r + r'c_1^2 + r(c_1')^2 \tag{3}$$

Now we shall use formula (1). Simple calculations lead to the relation

$$\chi(A \mid A') = 2rr' - c_1 \cdot c_1' - rc_2' - r'c_2 + \frac{r'c_1^2}{2} + \frac{r(c_1')^2}{2}. \tag{4}$$

Comparing formulæ (2), (3), (4), it is easy to convince oneself that the remark is true.

PROPOSITION 4. *Let S be a K3 surface and let $\pi : S \to \mathbf{P}^2$ be a double covering of $\mathbf{P}^2$ branched over a smooth curve of degree six. Let E be an exceptional bundle on $\mathbf{P}^2$. Then $\pi^*(E)$ is an exceptional bundle on S.*

PROOF. As was noted by S. Zube, exceptional bundles on $\mathbf{P}^2$, when restricted to a line ℓ, decompose into a direct sum $E|_\ell \cong n\mathcal{O}_\ell(d) \oplus k\mathcal{O}_\ell(d-1)$. Consequently, when $\pi^*(E)$ is restricted to $\ell' = \pi^{-1}(\ell)$, it also decomposes into a direct sum $\pi^*E|_{\ell'} \cong n\mathcal{O}_{\ell'}(d) \oplus k\mathcal{O}_{\ell'}(d-1)$, and therefore ${}^0\langle \pi^*E|_{\ell'} \mid \pi^*E|_{\ell'}(-3)\rangle = 0$. Thus ${}^0\langle \pi^*E \mid \pi^*E(-3)\rangle = 0$. From Lemma 3 we see that $\pi^*(E)$ is simple. Since $\chi(E, E) = \dim{}^0\langle E \mid E\rangle = 1$,

$$\chi(\pi^*E, \pi^*E) = 2 = \dim{}^0\langle \pi^*E \mid \pi^*E\rangle + \dim{}^1\langle \pi^*E \mid \pi^*E\rangle,$$

that is, π^*E is an exceptional bundle. The proposition is proved.

PROPOSITION 5. *Let (A, B) be an exceptional pair of bundles on $\mathbf{P}^2$. Then*

(1) (π^*A, π^*B) is an exceptional pair on S;

(2) ${}^0\langle A \mid B\rangle \cong {}^0\langle \pi^*A \mid \pi^*B\rangle$, that is, $\pi^*(L_A B) = L_{\pi^*A}\pi^*B$ and $\pi^*(R_A B) = R_{\pi^*A}\pi^*B$. That is, mutations commute with π^*.

PROOF. Any exceptional pair on $\mathbf{P}^2$ is contained in some helix. In other words, for an exceptional pair (A, B) there exists a bundle C such that the collection

$$\ldots A, B, C, A(3), B(3), C(3), \ldots \tag{$*$}$$

forms a helix ([1], Theorem 5.10). Note that if every exceptional pair in the helix $(*)$, lifts to an exceptional pair, then part (2) of the proposition will be satisfied for every exceptional pair in this helix $(*)$. Indeed, let (E, E') be some exceptional pair in the helix $(*)$. Since $(\pi^*E'(-3), \pi^*E)$ is also an exceptional pair, ${}^0\langle \pi^*E \mid \pi^*E'(-3)\rangle = 0$, but then ${}^0\langle \pi^*E \mid \pi^*E'\rangle \cong {}^0\langle E \mid E'\rangle$ from Lemma 3.

We can now prove the first part of the proposition by induction on mutations.

It is obvious that the proposition holds for line bundles. Let us assume that the proposition holds for every exceptional pair in the helix $(*)$. Let $E = \pi^*A$, $F = \pi^*B$, $G = \pi^*C$. We will consider the mutation

$$\cdots E(-3) \quad F(-3) \quad G(-3) \quad E \quad F \quad G \quad E(3) \quad F(3) \quad G(3) \cdots$$

$$\downarrow$$

$$\cdots F(-3) \quad E'(-3) \quad G(-3) \quad F \quad E' \quad G \quad F(3) \quad E'(3) \quad G(3) \cdots$$

Since the induction hypothesis implies that this mutation commutes with π^*, the bundle E' can be embedded into two exact sequences

$$0 \longrightarrow E \longrightarrow {}^0\langle E \mid F\rangle^* \otimes F \longrightarrow E' \longrightarrow 0 \qquad (**)$$

$$0 \longrightarrow E' \longrightarrow {}^0\langle G \mid E(3)\rangle \otimes G \longrightarrow E(3) \longrightarrow 0. \qquad (***)$$

We need to prove that (F, E'); $(G(-3), E')$; (E', G); $(E', F(3))$ are exceptional pairs, that is, the following relations are satisfied:

(1) $\quad {}^1\langle F \mid E'\rangle = {}^0\langle E' \mid F\rangle = 0,$

(2) $\quad {}^1\langle G(-3) \mid E'\rangle = {}^0\langle E' \mid G(-3)\rangle = 0,$

(3) $\quad {}^1\langle E' \mid G\rangle = {}^0\langle G \mid E'\rangle = 0,$

(4) $\quad {}^1\langle E' \mid F(3)\rangle = {}^0\langle F(3) \mid E'\rangle = 0.$

To prove the first two equalities, consider the exact sequence

$$\cdots \to {}^1\langle F \mid E(3)\rangle \to {}^2\langle F \mid E'\rangle \to {}^0\langle G \mid E(3)\rangle \otimes {}^2\langle F \mid G\rangle \to \cdots$$

From the induction hypothesis $(F, E(3))$ and (F, G) are exceptional pairs, therefore ${}^1\langle F \mid E(3)\rangle = {}^2\langle F \mid G\rangle = 0$, and consequently ${}^2\langle F \mid E'\rangle = {}^0\langle E' \mid F\rangle = 0$. Applying the functor $\langle F \mid$ to the exact sequence $(**)$ gives:

$$\cdots \to {}^0\langle E \mid F\rangle^* \otimes {}^1\langle F \mid F\rangle \to {}^1\langle F \mid E'\rangle \to$$

$$\to {}^1\langle F \mid E\rangle \xrightarrow{\phi} {}^0\langle E \mid F\rangle^* \otimes {}^2\langle F \mid F\rangle \to {}^2\langle F \mid E'\rangle \to 0.$$

Note that ${}^1\langle F \mid F\rangle = 0$, ${}^2\langle F \mid F\rangle \cong \mathbf{C}$ (since F is exceptional) and ϕ is an isomorphism, so that ${}^1\langle F \mid E'\rangle = {}^1\langle E' \mid F\rangle = 0$. The remaining relations can be proved in a similar way. The proposition is proved.

LEMMA 6. *Let $(\tilde{A}, \tilde{B})$ be an exceptional pair on $\mathbf{P}^2$, $A = \pi^* \tilde{A}$, $B = \pi^* \tilde{B}$. Then for arbitrary positive k*

(1) ${}^1\langle A \mid A(3k)\rangle = {}^1\langle A(3k) \mid A\rangle = 0,$

(2) ${}^1\langle A \mid B(3k)\rangle = {}^1\langle B(3k) \mid A\rangle = 0.$

PROOF. Since any exceptional bundle on $\mathbf{P}^2$ is contained in some helix, and mutations on $\mathbf{P}^2$ commute with π^*, the proof can be carried out by induction with respect to mutations on S. Let us consider the following mutation

$$\cdots A(-3) \quad B(-3) \quad C(-3) \quad A \quad B \quad C \cdots$$

$$\downarrow$$

$$\cdots B(-3) \quad A'(-3) \quad C(-3) \quad B \quad A' \quad C \cdots$$

$$0 \longrightarrow A \longrightarrow {}^0\langle A \mid B \rangle^* \otimes B \longrightarrow A' \longrightarrow 0 \qquad (*)$$

$$0 \longrightarrow A' \longrightarrow {}^0\langle C \mid A(3) \rangle \otimes C \longrightarrow A(3) \longrightarrow 0. \qquad (**)$$

We need to prove that, for $k > 0$,

(1) ${}^1\langle C \mid A'(3k) \rangle = 0$,

(2) ${}^1\langle A' \mid A'(3k) \rangle = 0$,

(3) ${}^1\langle A' \mid C(3k) \rangle = 0$,

(4) ${}^1\langle A' \mid B(3k) \rangle = 0$,

(5) ${}^1\langle B \mid A'(3k) \rangle = 0$.

We apply the functor $\langle C \mid$ to $(*)$ tensored with $\mathcal{O}(3k)$:

$${}^0\langle A \mid B \rangle^* \otimes {}^1\langle C \mid B(3k) \rangle \longrightarrow {}^1\langle C \mid A'(3k) \rangle \longrightarrow {}^2\langle C \mid A(3k) \rangle$$

$${}^1\langle C \mid A(3k) \rangle^* \cong {}^0\langle A \mid C(-3k) \rangle = 0, \qquad {}^1\langle C \mid B(3k) \rangle = 0.$$

Therefore ${}^1\langle C \mid A'(3k) \rangle = 0$.

We apply the functor $\langle A(3) \mid$ to $(*)$ tensored with $\mathcal{O}(3k)$:

$${}^0\langle A \mid B \rangle^* \otimes {}^1\langle A(3) \mid B(3k) \rangle \longrightarrow {}^2\langle A(k) \mid A'(3k) \rangle \longrightarrow 0,$$

$${}^2\langle A(3) \mid B(3k) \rangle^* \cong {}^0\langle B \mid A'(3 - 3k) \rangle \subset {}^0\langle B \mid A \rangle = 0.$$

Therefore ${}^2\langle A(3) \mid A'(3k) \rangle = 0$.

We apply the functor $\mid A'(3k) \rangle$ to $(**)$:

$${}^0\langle C \mid A(3) \rangle^* \otimes {}^1\langle C \mid A'(3k) \rangle \longrightarrow {}^1\langle A' \mid A'(3k) \rangle \longrightarrow {}^2\langle A(3) \mid A'(3k) \rangle.$$

From what we proved above we see that ${}^1\langle A \mid A'(3k) \rangle = 0$. The remaining properties can be proved in a similar way.

THEOREM 7. *Let S be a K3 surface which is a double cover of $\mathbf{P}^2$ branched over a smooth curve of degree six. Suppose that the helix $\ldots A_0, A_1, A_2, A_3, A_4, A_5, \ldots$ on S is obtained from the helix $\ldots \mathcal{O}_S, \mathcal{O}_S(1), \mathcal{O}_S(2), \mathcal{O}_S(3), \ldots$ by mutations. Then for any positive k*

(1) (A_0, A_k) is an exceptional pair.

(2) The canonical map ${}^0\langle A_0 \mid A_k \rangle \otimes A_0 \to A_k$ is an epimorphism.

(3) The canonical map $A_0 \to {}^0\langle A_0 \mid A_k \rangle^* \otimes A_k$ is a monomorphism.

PROOF. (1) Since A_0 and A_k are in the same helix, ${}^1\langle A_0 \mid A_k \rangle = 0$; moreover either $A_k = A_0(3n)$, or $A_k = B(3n)$, where (A_0, B) is an exceptional pair. In the first case ${}^1\langle A_0 \mid A_0(3n) \rangle^* \cong {}^0\langle A_0 \mid A_0(-3n) \rangle = 0$, because A_0 is a simple bundle. In the second case ${}^1\langle A_0 \mid B(3n) \rangle^* \cong {}^0\langle B \mid A_0(-3n) \rangle \subset {}^0\langle B \mid A \rangle = 0$, that is, $(A_0, A - K)$ is an exceptional pair.

Parts (2) and (3) of the theorem follow from Lemma 2 and Proposition 5.

4. Exceptional Bundles on a Quartic.

In this section, we study exceptional bundles on a K3 hypersurface in $\mathbf{P}^3$. The idea of studying such a surface in this context is due to M. Karpanov.

THEOREM 8. *Let X be a hypersurface of degree four in $\mathbf{P}^3$, E an exceptional bundle on $\mathbf{P}^3$, (A, B) an exceptional pair on $\mathbf{P}^3$ and $\ldots A_1, A_2, A_3, A_4, \ldots$ a constructable helix on $\mathbf{P}^3$. Then*

(1) $E|_X$ is an exceptional bundle on X.

(2) $(A|_X, B|_X)$ is an exceptional pair.

(3) $^0\langle A \mid B \rangle \cong {}^0\langle A|_X \mid B|_X \rangle$.

(4) For any integers k and n such that $k < n$, $(A_k|_X, A_n|_X)$ is an exceptional pair on X, and the canonical sequences

$$^0\langle A_k|_X \mid A_n|_X \rangle \otimes A_k|_X \longrightarrow A_n|_X \longrightarrow 0$$

$$0 \longrightarrow A_k|_X \longrightarrow {}^0\langle A_k|_X \mid A_n|_X \rangle^* \otimes A_n|_X$$

are exact.

PROOF. Consider the long exact sequence in cohomology of the following short exact sequence

$$0 \longrightarrow E^* \otimes E(-4) \longrightarrow E^* \otimes E \longrightarrow E^* \otimes E|_X \longrightarrow 0 :$$

i.e.

$$0 \to H^0\big(E^* \otimes E(-4)\big) \to H^0(E^* \otimes E) \to H^0(E \otimes E|_X) \to$$
$$\to H^1\big(E^* \otimes E(-4)\big) \to H^1(E^* \otimes E) \to H^1(E^* \otimes E|_X) \to H^2\big(E^* \otimes E(-4)\big)$$

Serre duality implies $H^i\big(E^* \otimes E(-4)\big)^* \cong H^{3-i}(E^* \otimes E)$ and since E is an exceptional bundle, $H^i\big(E^* \otimes E(-4)\big) = 0$, for $i = 0, 1, 2$. Since $H^0(E^* \otimes E) \cong \mathbf{C}$, $H^1(E^* \otimes E) = 0$, $E|_X$ is an exceptional bundle.

From the structure sequence tensored with $A^* \otimes B$ we get the following long exact sequence:

$$0 \to {}^0\langle A \mid B(-4) \rangle \to {}^0\langle A \mid B \rangle \to {}^0\langle A|_X \mid B|_X \rangle \to {}^1\langle A \mid B(-4) \rangle \to$$
$$\to {}^1\langle A \mid B \rangle \to {}^1\langle A|_X \mid B|_X \rangle \to {}^2\langle A \mid B(-4) \rangle \to \qquad (*)$$
$$\to {}^2\langle A \mid B \rangle \to {}^2\langle A|_X \mid B|_X \rangle \to {}^3\langle A \mid B(-4) \rangle.$$

From Serre duality $H^i(A^* \otimes B)^* \cong H^{3-i}\big(A^* \otimes B(-4)\big)$. (A, B) is an exceptional pair on $\mathbf{P}^3$ and consequently $H^i\big(A^* \otimes B(-4)\big) = 0$ for all i and $H^i(A^* \otimes B) = 0$ for $i \neq 0$.

Now, from sequnece (*) it follows immediately, that the pair $(A|_X, B|_X)$ is exceptional and $^0\langle A \mid B \rangle \cong {}^0\langle A|_X \mid B|_X \rangle$.

From the part of theorem already proved, it follows that the restriction of a helix to X remains a helix, but A_{k+s} and A_k differ, not by twisting with the canonical class, but by $\mathcal{O}(-4)$. Therefore the proof of the fourth statement can be carried out by induction with respect to mutations, noting that a helix composed of line bundles satisfies the theorem.

Consider the mutation on X:

$$A, B, C, D, A(4)$$

$$\downarrow$$

$$B, A', C, D, B(4)$$

If we set $^0\langle A \mid B \rangle = V$, $^0\langle D \mid A(4) \rangle = W$, $^0\langle C \mid \tilde{A} \rangle = L$, then we can write the following sequences

$$0 \longrightarrow A \longrightarrow V^* \otimes B \longrightarrow A' \longrightarrow 0 \tag{5}$$

$$0 \longrightarrow \tilde{A} \longrightarrow W \otimes D \longrightarrow A(4) \longrightarrow 0 \tag{6}$$

$$0 \longrightarrow A' \longrightarrow L \otimes C \longrightarrow \tilde{A} \longrightarrow 0 \tag{7}$$

We need to check that, for arbitrary positive k,

(a) $^1\langle B \mid A'(4k) \rangle = 0$,   (e) $^1\langle A' \mid B(4k) \rangle = 0$,

(b) $^1\langle C \mid A'(4k) \rangle = 0$,   (f) $^1\langle A' \mid D(4k) \rangle = 0$,

(c) $^1\langle D \mid A'(4k) \rangle = 0$,   (g) $^1\langle A' \mid A'(4k) \rangle = 0$.

(d) $^1\langle A' \mid C(4k) \rangle = 0$,

First of all note that $^2\langle A_k|_X \mid A_n|_X \rangle = 0$. Indeed, since A_k and A_n are in the same helix, either $A_n = A_k(4s)$, or $A_n = B(4s)$, where A_k and B form an exceptional pair. Then

$$\mathrm{Ext}^2(A_k|_X, A_k(4s)|_X)^* \cong \mathrm{Hom}(A_k|_X, A_k(-4s)|_X) = 0$$

$$\mathrm{Ext}^2(A_k|_X, B(4s)|_X) \cong \mathrm{Hom}(B|_X, A_k(-4s)|_X) = 0$$

(a) Apply the functor $\langle B \mid$ to sequence (5) tensored with $\mathcal{O}_X(4n)$:

$$V^* \otimes {}^1\langle B \mid B(4k) \rangle \longrightarrow {}^1\langle B \mid A'(4k) \rangle \longrightarrow {}^2\langle B \mid A(4k) \rangle.$$

$^1\langle B \mid B(4k) \rangle = 0$ from the induction hypothesis and $^2\langle B \mid A(4k) \rangle = 0$ from the remark. Consequently $^1\langle B \mid A'(4k) \rangle = 0$. We can prove properties (b) and (c) similarly.

(d) Applying the functors $|C(4k)\rangle$, $|B(4k)\rangle$, $|D(4k)\rangle$ to sequence (6) we find that $^2\langle \tilde{A} \mid C(4k)\rangle = {}^2\langle \tilde{A} \mid D(4k)\rangle = 0$, $^2\langle \tilde{A} \mid B(4k)\rangle = 0$. Apply the functor $|C(4k)\rangle$ to sequence (7):

$$L^* \otimes {}^1\langle C \mid C(4k)\rangle \longrightarrow {}^1\langle A' \mid C(4k)\rangle \longrightarrow {}^2\langle \tilde{A} \mid C(4k)\rangle$$

from which it follows that $^1\langle A' \mid C(4k)\rangle = 0$. Parts (e) and (f) can be proved similarly.

Applying the functor $|A'(4k)\rangle$ to sequence (6) we get $^2\langle \tilde{A} \mid A'(4k)\rangle = 0$ for $k > 0$. But then the last part (g) follows from the sequence

$$L^* \otimes {}^1\langle C \mid A'(4k)\rangle \longrightarrow {}^1\langle A' \mid A'(4k)\rangle \longrightarrow {}^2\langle \tilde{A} \mid A'(4k)\rangle.$$

In order to prove that the canonical maps are epimorphic or monomorphic, we need to apply Lemma 2. This completes the proof of the theorem.

References

[1] GORODENTSEV, A.L. & RUDAKOV, A.N., Exceptional Vector Bundles on Projective Space, *Duke Math. J.*, **54** (1987) 115–130.

[2] MUKAI, S., On the Moduli Spaces of Bundles on K3 Surfaces, I, in *Vector Bundles* ed. Atiyah et al, Oxford Univ. Press, Bombay, (1986) 341–413.

[3] RUDAKOV, A.N., The Markov Numbers and Exceptional Bundles on $\mathbf{P}^2$, *Math. USSR Isv.*, **32** (1989) 99–112.

11. The Stability of Exceptional Bundles on Three Dimensional Projective Space

D. Yu. Zube

In this paper we establish the μ-stability of exceptional bundles on $\mathbf{P}^3$. The basic idea of the proof is to restrict the given bundle to a K3 surface in $\mathbf{P}^3$ with Picard group $\mathbf{Z}$, in other words to a hypersurface of degree 4. The restriction is seen to be an exceptional bundle over this surface. But then, using some results of Mukai we can show that such bundles are μ-stable in the sense of Mumford-Takemoto with respect to a polarization given by any ample divisor.

THEOREM. *Let E be an exceptional bundle on a K3 surface X with Picard group $\mathbf{Z}$. Then E is μ-stable with respect to any ample divisor H.*

PROOF. We start by showing that E is μ-semistable. If not, then there exists a short exact sequence

$$0 \to F \to E \to G \to 0 \tag{1}$$

where F and G are torsion free sheaves and G can be chosen to be μ-stable. However,

$$\mu_H(F) = \frac{c_1(F)}{r(F)} \cdot H \geqslant \mu_H(E) > \mu_H(G).$$

This implies that $\mathrm{Hom}(F, G) = 0$ and, from Mukai [1], we have

$$\dim {}^1\langle E \mid E \rangle \geqslant \dim {}^1\langle F \mid F \rangle + \dim {}^1\langle G \mid G \rangle,$$

ie. $${}^1\langle F \mid F \rangle = {}^1\langle G \mid G \rangle = 0, \quad \chi(F, F) \geqslant 2, \quad \chi(G, G) \geqslant 2.$$

Mukai also shows in [1] that for the bundles in (1)

$$\frac{\chi(E, E)}{r(E)} - \frac{\chi(F, F)}{r(F)} - \frac{\chi(G, G)}{r(G)} = \frac{r(F)r(G)}{r(E)}\bigl(\mu(F) - \mu(G)\bigr)^2 \tag{2}$$

From the fact that $\mathrm{Pic}\, X = \mathbf{Z}$, it follows that the RHS of (2) is positive. But this contradicts the inequalities $\chi(F, F) \geqslant 2$, $\chi(G, G) \geqslant 2$. This means that there are no such F, and so E is μ-semistable.

Suppose that E is not μ-stable, then there is a sequence (1) such that $\mu_H(F) = \mu_H(E) = \mu_H(G)$ and hence $\bigl(\mu(F) - \mu(E)\bigr) \cdot H = 0$. Then from the Hodge Index Theorem we have $\bigl(\mu(F) - \mu(E)\bigr)^2 \leqslant 0$ and, since $\mathrm{Pic}\, X = \mathbf{Z}$, $\mu(E) = \mu(F)$. Now put

$$\mu(E) = \frac{m(E).A}{r(E)}$$

where $m(E) \in \mathbf{Z}$ and A is a generator of the Picard group. If we can show that $m(E)$ and $r(E)$ are coprime, then there can be no numbers m(F) and r(F) such that

$$\frac{m(F)}{r(F)} = \frac{m(E)}{r(E)},$$

where $r(F) < r(E)$, and hence semistability and stability will coincide. Let $c_1(E) = c_1$, $c_2(E) = c_2$, $m(E) = m$, $r(E) = r$, then

$$\chi(E, E) = \left((r, -c_1, \tfrac{1}{2}c_1^2 - c_2)(r, c_1, \tfrac{1}{2}c_1^2 - c_2)(1, 0, 2) \right),$$

where $(1, 0, 2) = \operatorname{Td} X$ is the Todd class, and since $\chi(E, E) = 2$,

$$2r^2 - c_1^2 + 2r \left(\tfrac{1}{2}c_1^2 - c_2 \right) = 2.$$

Substituting $m.A$ for c_1, we have

$$2r^2 - m^2 A^2 + m^2 A^2 r - 2rc_2 = 2.$$

From this we see that m and r are coprime, as required.

COROLLARY. *If E is an exceptional bundle on $\mathbf{P}^3$, then E is μ-stable.*

PROOF. Suppose that E is not μ-stable then there exists a torsion free subsheaf F of E such that $\mu(F) \geqslant \mu(E)$. Let X be a K3 surface in $\mathbf{P}^3$ such that $\operatorname{Pic} X = \mathbf{Z}$. Then $F|_X$ will be torsion free, whenever the singularity set $S(F)$ intersects X in points. Since $S(F)$ is at most a curve we can replace F by g^*F for some $g \in \operatorname{Aut} \mathbf{P}^3$ to remove the singularities. However E is homogeneous so that $E \cong g^*E$ and hence we can assume that $F|_X$ is a subbundle of $E|_X$.

Furthermore, S. Kulishov has shown that $E|_X$ is an exceptional bundle over X. The theorem now implies that $E|_X$ is μ-stable with respect to any hyperplane section H. But then

$$\mu_H(F|_X) = \frac{c_1(F) \cdot H}{r(F)} < \frac{c_1(E) \cdot H}{r(E)} = \mu_H(E|_X),$$

which contradicts the assumption.

REMARK. To show μ-stability it would be enough to know that $E|_X$ is μ-semistable from which we could deduce that E is μ-stable on $\mathbf{P}^3$. This could be done, as in the second part of the proof of the theorem, using Riemann-Roch to show that the condition $\chi(E, E) = 2$ implies that r and c_1 are coprime, which would prove that stability and semistability for E are equivalent.

On $\mathbf{P}^2$, $\mathbf{P}^1 \times \mathbf{P}^1$ and K3 surfaces, since exceptional bundles are stable, they are uniquely determined by their rank and first Chern class. On $\mathbf{P}^3$ we cannot have this. It is not known whether an exceptional bundle on $\mathbf{P}^3$ is determined by its total Chern class.

References

[1] MUKAI, S., On the Moduli Spaces of Bundles on K3 Surfaces, I, in *Vector Bundles* ed. Atiyah et al, Oxford Univ. Press, Bombay, (1986) 341–413.

12. A Symmetric Helix on the Plücker Quadric

B.V. Karpov

In this paper we construct a helix composed of exceptional bundles on the complex Grassmannian $G(2,4)$. We shall identify $G(2,4)$ with its image G under the Plücker embedding, which is a quadric in $\mathbf{P}^5$. The constructed helix is in some ways similar to the one on $\mathbf{P}^1 \times \mathbf{P}^1$ (see [3] or the papers in this collection): its foundation consists of spinor bundles and line bundles. In a more general setting, such a collection of bundles is considered in [2], where it is used to study the derived category of coherent sheaves on a quadric in $\mathbf{P}^n$. There, in particular, it is shown that there are no higher Ext's between the bundles in this collection. In order to give a complete presentation here, we include an elementary proof of this fact for the Plücker quadric.

From the helix we construct, we can obtain a countable set of exceptional bundles on G. However, we shall not study this set here. The question of how it relates to the set of all exceptional bundles on G is still open.

As a secondary result, we obtain the long exact sequences (3.2.15) and (3.3.4) of vector bundles on G, which are analogues of the Koszul complex on $\mathbf{P}^5$. It is a conjecture that similar sequences can be constructed on a quadric in $\mathbf{P}^n$ for any n.

1. General Notions.

1.1. Definition.

A vector bundle E on G is *exceptional* if

1) $^0\langle E \mid E \rangle \cong \mathbf{C}$.

2) $^i\langle E \mid E \rangle = 0$ for $i \geqslant 1$.

Note that, if the bundle E is exceptional, then the bundles E^* and $E \otimes L$ are also exceptional, for an arbitrary line bundle L.

1.2. Definition.

An ordered collection $(E_1, \ldots, E_n)$ of vector bundles on G is *exceptional* if

1) E_m is exceptional, for $m = 0, \ldots, n$.

2) $^i\langle E_l \mid E_m \rangle = 0$ for $i \geqslant 1$ and $l < m$.

3) $^i\langle E_m \mid E_l \rangle = 0$ for $i \geqslant 0$ and $l < m$.

Such collections of bundles arise on projective spaces (see [1]). However, on slightly more complicated varieties it is convenient to consider collections in which some bundle E_m is replaced by an unordered pair of bundles (F_1, F_2). Such pairs will be written in a column.

1.3. Definition.

The collection of exceptional bundles on G

$$\sigma = \left(E_1, \ldots, E_{m-1}, \frac{F_1}{F_2}, E_{m+1}, \ldots, E_n\right)$$

is *exceptional* if

1) The collections $(E_1, \ldots, E_{m-1}, F_j, E_{m+1}, \ldots, E_n)$, for $j = 1, 2$, are exceptional in the sense of definition 1.2.

2) $^i\langle F_j \mid F_{3-j}\rangle = 0$ for $j = 1, 2$ and $i \geqslant 0$. Note that, in this case, the two collections

$$\sigma^* = \left(E_n^*, \ldots, E_{m+1}^*, \frac{F_1^*}{F_2^*}, E_{m-1}^*, \ldots, E_1^*\right)$$

$$\sigma \otimes L = \left(E_1 \otimes L, \ldots, E_{m-1} \otimes L, \frac{F_1 \otimes L}{F_2 \otimes L}, E_{m+1} \otimes L, \ldots, E_n \otimes L\right)$$

are also exceptional, for any line bundle L.

1.4. Mutations of elementary collections.

Let $\sigma = (A, B)$ be an ordered pair of bundles. We will assume that there exists one of the following exact sequences of bundles on G:

$$0 \to B' \to A \otimes {}^0\langle A \mid B\rangle \xrightarrow{\text{can}} B \to 0 \tag{1.4.1}$$

$$0 \to A \xrightarrow{\text{can}} {}^0\langle A \mid B\rangle^* \otimes B \to A' \to 0, \tag{1.4.2}$$

where $\text{can} \in H^0\left(A^* \otimes {}^0\langle A \mid B\rangle^* \otimes B\right) \cong {}^0\langle A \mid B\rangle^* \otimes {}^0\langle A \mid B\rangle$ is induced by the identity endomorphism of the space ${}^0\langle A \mid B\rangle$. In the case when (1.4.1) exists we say that B' is the *left shift* of B in σ and B' is denoted by $L_\sigma B$. Sometimes we will say that the pair of bundles $L(A, B) = (B', A)$ is a *left mutation* of the pair (A, B). A similar notation is used in the case of (1.4.2): $A' = R_\sigma A$ is called the *right shift* of A in the pair σ and $R\sigma = (B, A')$ is called the *right mutation* of σ.

Let $\sigma = \left(\frac{A_1}{A_2}, B\right)$ be collection of bundles. We will assume that there exists the following exact sequence of bundles on G:

$$0 \to B' \to \begin{matrix} A_1 \otimes {}^0\langle A_1 \mid B\rangle \\ \oplus \\ A_2 \otimes {}^0\langle A_2 \mid B\rangle \end{matrix} \xrightarrow{\text{can}} B \to 0, \tag{1.4.3}$$

where

$$can \in H^0\left(\left(A_1^* \otimes {}^0\langle A_1 \mid B\rangle^* \quad \oplus \quad A_2^* \otimes {}^0\langle A_2 \mid B\rangle^*\right) \otimes B\right)$$

$$\cong {}^0\langle A_1 \mid B\rangle^* \otimes {}^0\langle A_1 \mid B\rangle \quad \oplus \quad {}^0\langle A_2 \mid B\rangle^* \otimes {}^0\langle A_2 \mid B\rangle$$

is induced by the pair of identity endomorphisms of the spaces

$$^0\langle A_1 \mid B\rangle \quad \text{and} \quad ^0\langle A_2 \mid B\rangle.$$

In this case we say that $B' = L_\sigma B$ is the *left shift* of B in σ and the collection $L\sigma = \left(B', \begin{smallmatrix}A_1\\A_2\end{smallmatrix}\right)$ is called the *left mutation* of σ. If $R(A_1, B) = (B, A_1')$ and $R(A_2, B) = (B, A_2')$ are defined, then the *right mutation* of σ is, by definition, the collection $R\sigma = \left(B, \begin{smallmatrix}A_1'\\A_2'\end{smallmatrix}\right)$ and the set $R\left(\begin{smallmatrix}A_1\\A_2\end{smallmatrix}\right) = \left(\begin{smallmatrix}A_1'\\A_2'\end{smallmatrix}\right)$ is called the *right shift* of $\left(\begin{smallmatrix}A_1\\A_2\end{smallmatrix}\right)$ in σ.

Now let $\sigma = \left(A, \begin{smallmatrix}B_1\\B_2\end{smallmatrix}\right)$. We will assume that there exists a short exact sequence of bundles on G

$$0 \to A \xrightarrow{\text{can}} \begin{matrix} ^0\langle A \mid B_1\rangle^* \otimes B_1 \\ \oplus \\ ^0\langle A \mid B_2\rangle^* \otimes B_2 \end{matrix} \to A' \to 0, \qquad (1.4.4)$$

where

$$\text{can} \in H^0\left(A^* \otimes \left(^0\langle A \mid B_1\rangle^* \otimes B_1 \oplus {}^0\langle A \mid B_2\rangle^* \otimes B_2\right)\right)$$
$$\cong {}^0\langle A \mid B_1\rangle^* \otimes {}^0\langle A \mid B_1\rangle \oplus {}^0\langle A \mid B_2\rangle^* \otimes {}^0\langle A \mid B_2\rangle$$

is induced by the pair of identity endomorphisms of the spaces

$$^0\langle A \mid B_1\rangle \quad \text{and} \quad ^0\langle A \mid B_2\rangle.$$

In this case we say that $A' = R_\sigma$ is the *right shift* of A in σ and the mutation $R\sigma = \left(\begin{smallmatrix}B_1\\B_2\end{smallmatrix}, A'\right)$ is called the *right mutation* of σ. If $L(A, B_1) = (B_1', A)$ and $L(A, B_2) = (B_2', A)$ are defined then the *left mutation* of σ is, by definition, $L\sigma = \left(\begin{smallmatrix}B_1'\\B_2'\end{smallmatrix}, A\right)$ and the set $L_\sigma\left(\begin{smallmatrix}B_1\\B_2\end{smallmatrix}\right) = \left(\begin{smallmatrix}B_1'\\B_2'\end{smallmatrix}\right)$ is called the *left shift* of $\left(\begin{smallmatrix}B_1\\B_2\end{smallmatrix}\right)$ in σ.

1.5. Remark.

Let $E = E_1 \oplus E_2$, where E_1 and E_2 are exceptional bundles such that

$$^i\langle E_j \mid E_{3-j}\rangle = 0, \quad \text{for } i \geqslant 0 \text{ and } j = 1, 2.$$

Then E_1 and E_2 can be uniquely recovered from E. Indeed, the space

$$\text{Hom}(E, E) \cong \text{Hom}(E_1, E_1) \oplus \text{Hom}(E_2, E_2) \cong \mathbb{C} \oplus \mathbb{C}$$

has a $\mathbb{C}$-algebra structure such that $1 = p_1 + p_2$, where $p_i^2 = p_i$ and $p_i p_{3-i} = 0$ for $i = 1, 2$. The E_i are obtained as the images of the p_i. The fact that the E_i are uniquely determined follows from the uniqueness of the decomposition of the unit element in the algebra $\text{Hom}(E, E)$ into a sum of orthogonal idempotents. This remark allows us later on to identify collections of bundles of the form $\left(E, \begin{smallmatrix}F_1\\F_2\end{smallmatrix}\right)$ and $\left(\begin{smallmatrix}E_1\\E_2\end{smallmatrix}, F\right)$ with pairs of bundles $(E, F_1 \oplus F_2)$ and $(E_1 \oplus E_2, F)$ respectively.

1.6. Definition.

1)   An exceptional pair of bundles $\sigma = (A, B)$ is called *left admissable* (resp. *right admissable*) if a left (resp. right) mutation is possible in it. We will also write $L_A B$ for $L_\sigma B$ (resp. $R_B A = R_\sigma A$).

2)   A pair of bundles $(A_1 \oplus A_2, B)$, identified with the exceptional collection $\sigma = \left({A_1 \atop A_2}, B \right)$, is called *left admissable* (resp. *right admissable*) if a left (resp. right) mutation is possible in σ. We will also write $L_{A_1 \oplus A_2} B$ for $L_\sigma B$ (resp. $R_\sigma A_1 \oplus A_2 = A_1' \oplus A_2'$ where $R(A_i, B) = (B, A_i')$ for $i = 1, 2$).

3)   The pair of bundles $(A, B_1 \oplus B_2)$, identified with the exceptional collection $\sigma - \left(A, {B_1 \atop B_2} \right)$, is called *right admissable* (resp. *left admissable*), if a right (resp. left) mutation is possible in σ. We will also write $R_{B_1 \oplus B_2} A$ for $R_\sigma A$ (resp. $L_A B_1 \oplus B_2 = B_1' \oplus B_2'$ where $L(A, B_i) = (B_i', A)$ for $i = 1, 2$).

Later on, when we write (A, B) for an admissable pair, we will mean one of the three cases listed above.

1.7. Proposition.

(1) If the pair (A, B) is left admissable, then the pair $(L_A B, A)$ is right admissable and $R_A L_A B = B$.

(2) If the pair (A, B) is right admissable, then the pair $(B, R_B A)$ is left admissable and $L_B R_B A = A$.

PROOF. 1) The case when (A, B) is an exceptional pair is analysed in [**1**, 1.5,1.6].

2)   Let $(A, B) = (A_1 \oplus A_2, B)$ be identified with the exceptional collection $\left({A_1 \atop A_2}, B \right)$. Applying the functor $\langle A_j \mid$, with $j = 1, 2$, to (1.4.3), we get ${}^i \langle A_j \mid B' \rangle = 0$, for $i \geqslant 0$. Applying the functor $\mid A_j \rangle$ to (1.4.3), we deduce that ${}^i \langle B' \mid A_j \rangle = 0$, for $i \geqslant 1$, and that there exist canonical isomorphisms ${}^0 \langle B' \mid A_j \rangle \cong {}^0 \langle A_j \mid B \rangle^*$, for $j = 1, 2$. Therefore, we can rewrite (1.4.3) in the form

$$0 \to B' \to \begin{matrix} A_1 \otimes {}^0 \langle B' \mid A_1 \rangle^* \\ \oplus \\ A_2 \otimes {}^0 \langle B' \mid A_2 \rangle^* \end{matrix} \to B \to 0,$$

which proves (1) in this case. Applying first the functor $\mid A_{3-j} \rangle$, and then $\langle R_B A_{3-j} \mid$, to the exact sequences defining $R_B A_j$, we obtain first ${}^i \langle R_B A_j \mid A_{3-j} \rangle = 0$, for $i \geqslant 0$, and then ${}^i \langle R_B A_{3-j} \mid R_B A_j \rangle = 0$, for $i \geqslant 0$. This, together with [**1**, 1.5,1.6], completes the proof of (2).

3)   The case $(A, B) = (A, B_1 \oplus B_2)$ is similar to (2).

This proposition means that the first axiom of admissable pairs (see Paper 1 in this collection) is true. The second axiom can be checked just as in the proof of Lemma 2.2 in [1]. Our case is only slightly more complicated, as partially demonstrated in the above proof.

The helices of vector bundles on G, which are obtained in the manner given in Proposition 1.6 of paper 5 in this collection, are analogous to the helices in [1]. The remainder of this paper will be devoted to constructing a helix of period 5 on G, which, we hope, will help to describe the exceptional bundles on G.

2.   Exceptional Collections on G.

Let $G(2, 4) = G(2, V)$, where $\dim V = 4$. Let us write $W = \Lambda^2 V$, so that $\dim W = 6$ and G is a quadric in $\mathbf{P}(W)$.

2.1. Examples of Exceptional Bundles on G.

The line bundles on G take the form $\mathcal{O}_G(d) = \mathcal{O}_{\mathbf{P}^5}(d)|_G$. It is clear that these are all exceptional.

We will denote by U the universal subbundle over G and by Q the universal quotient bundle. We have $\operatorname{rk} U = \operatorname{rk} Q = 2$ and $\det U = \det Q^* = \mathcal{O}_G(-1)$. It is well-known that, for any two-dimensional bundle E over any variety, $E \cong E^* \otimes \det E$. Thus we get $U = U^*(-1)$ and $Q = Q^*(1)$.

Let $\mathbf{P}^3 = \mathbf{P}(V)$. Then G can be interpreted as the set of lines in $\mathbf{P}^3$. The flag variety $F = \{(p, \ell) \in \mathbf{P}^3 \times G \mid p \in \ell\}$ has two projections

$$F$$
$$\pi_0 \swarrow \qquad \searrow \pi_1$$
$$\mathbf{P}^3 \qquad\qquad G$$

The fibre of the bundle $\mathbf{R}^0 \pi_{1*} \pi_0^* \mathcal{O}_{\mathbf{P}^3}(1)$ over a point $\ell \in G$ coincides with the fibre U^* over ℓ. If $\ell = \mathbf{P}(L)$ for $L \subset V$, then

$$H^0\big(\mathcal{O}_{\pi_1^{-1}(\ell)}(1)\big) = H^0\big(\mathcal{O}_\ell(1)\big) = L^* = U_\ell^*$$

This implies that $U^* = \mathbf{R}^0 \pi_{1*} \pi_0^* \mathcal{O}_{\mathbf{P}^3}(1)$. In the same way, $S^2 U^* = \mathbf{R}^0 \pi_{1*} \pi_0^* \mathcal{O}_{\mathbf{P}^3}(2)$. From the projection formula we obtain

$$U = U^*(-1) = \mathbf{R}^0 \pi_{1*}\big(\pi_0^* \mathcal{O}_{\mathbf{P}^3}(1) \otimes \pi_1^* \mathcal{O}_G(-1)\big),$$
$$S^2 U^*(d) = \mathbf{R}^0 \pi_{1*}\big(\pi_0^* \mathcal{O}_{\mathbf{P}^3}(2) \otimes \pi_1^* \mathcal{O}_G(d)\big).$$

Using this construction, we deduce the following:

$$H^i(U^*) = \begin{cases} V^* & \text{for } i = 0 \\ 0 & \text{for } i \geqslant 1 \end{cases} \qquad H^i(U) = 0, \quad \text{for } i \geqslant 0 \qquad (2.1.1)$$

$$H^i\big(S^2 U^*(-1)\big) = 0, \quad \text{for } i \geqslant 0, \qquad\qquad (2.1.2)$$

$$H^i\big(S^2 U^*(-2)\big) = 0, \quad \text{for } i \geqslant 0. \qquad\qquad (2.1.3)$$

The fact that U is exceptional follows from (2.1.2) and from the formula

$$U^* \otimes U = U^* \otimes U^* \otimes \mathcal{O}_G(-1) = \big(S^2 U^* \oplus \Lambda^2 U^*\big) \otimes \mathcal{O}_G(-1) = S^2 U^*(-1) \oplus \mathcal{O}_G$$

We deduce that Q is exceptional from the fact that U is exceptional and the short exact sequences

$$0 \to U \to V \otimes \mathcal{O}_G \to Q \to 0, \qquad\qquad (2.1.4)$$

$$0 \to Q^* \to V^* \otimes \mathcal{O}_G \to U^* \to 0. \qquad\qquad (2.1.5)$$

Now, $H^i(Q^*) = 0$, for $i \geqslant 0$, because the map $V^* \otimes \mathcal{O}_G \to U^*$ from (2.1.5) induces an isomorphism on sections.

Henceforth, we will denote the exact sequence $(k.l.m)$, tensored by the bundle E, by $(k.l.m) \otimes E$ and the sequence dual to $(k.l.m)$ by $(k.l.m)^*$.

From (2.1.5) $\otimes U$ and (2.1.1) we find that $H^1(Q^* \otimes U) = \mathbf{C}$ and $H^i(Q^* \otimes U) = 0$ for $i \neq 1$. From (2.1.4) $\otimes Q^*$ we see that $H^0(Q^* \otimes Q) = \mathbf{C}$ and $H^i(Q^* \otimes Q) = 0$ for $i \geqslant 1$, i.e. Q is exceptional.

2.2. Tables of Dimensions for $H^i\big(U^*(d)\big)$, $H^i\big(Q^*(d)\big)$ and $H^i\big(\mathcal{O}_G(d)\big)$.

Since G is a quadric in $\mathbf{P}^5$, we can use the exact sequence

$$0 \to \mathcal{O}_{\mathbf{P}^5}(-2) \to \mathcal{O}_{\mathbf{P}^5} \to \mathcal{O}_G \to 0$$

to compute $h^i\big(\mathcal{O}_G(d)\big)$ for all i and d:

$i\backslash d$	$\cdots$	-6	-5	-4	-3	-2	-1	0	1	2	3	4	$\cdots$
0	$\cdots$	0	0	0	0	0	0	1	6	20	50	105	$\cdots$
1	$\cdots$	0	0	0	0	0	0	0	0	0	0	0	$\cdots$
2	$\cdots$	0	0	0	0	0	0	0	0	0	0	0	$\cdots$
3	$\cdots$	0	0	0	0	0	0	0	0	0	0	0	$\cdots$
4	$\cdots$	20	6	1	0	0	0	0	0	0	0	0	$\cdots$

Indeed, for $d \geqslant 0$, we have

$$h^0(\mathcal{O}_G(d)) = \frac{(d+1)(d+2)^2(d+3)}{12},$$
$$h^i(\mathcal{O}_G(d)) = 0 \quad \text{for} \quad i \geqslant 1.$$

Serre duality and the fact that the canonical class $K_G = \mathcal{O}_G(-4)$ gives the other set of nonzero entries.

Below we give the table of $h^i(U^*(d))$ and $h^i(Q(d))$. It is not difficult to find these dimensions. If we know $h^i(U(d))$, then from (2.1.4) $\otimes \mathcal{O}_G(d+1)$ we can compute $h^i(Q(d)) = h^i(Q^*(d+1))$ and then from (2.1.5) $\otimes \mathcal{O}_G(d+1)$ we find $h^i(U(d+2))$. This process can be repeated as often as necessary. If we start with $d = 0$ (c.f. (2.1.1)), we can find $h^i(U(2k))$ and $h^i(Q(2k))$, whereas, if we start with $d = 1$, we find

$$h^i(U(2k+1)) \quad \text{and} \quad h^i(Q(2k+1)), \quad \text{for } k \geqslant 0.$$

We proceed similarly for negative d.

As a result of all these calculations we have $h^i(U^*(d)) = h^i(Q(d))$, and the table of values of $h^i(U^*(d))$ is

$i \backslash d$		-6	-5	-4	-3	-2	-1	0	1	2	3	4	
0	$\cdots$	0	0	0	0	0	0	4	20	60	140	280	$\cdots$
1	$\cdots$	0	0	0	0	0	0	0	0	0	0	0	$\cdots$
2	$\cdots$	0	0	0	0	0	0	0	0	0	0	0	$\cdots$
3	$\cdots$	0	0	0	0	0	0	0	0	0	0	0	$\cdots$
4	$\cdots$	20	4	0	0	0	0	0	0	0	0	0	$\cdots$

It is interesting to note that, by Serre duality, we have

$$h^i(U^*(d)) = h^{4-i}(U(-d-4)) = h^{4-i}(U^*(-d-5))$$

and similarly $h^i(Q(d)) = h^{4-i}(Q(-d-5))$.

2.3. Proposition.

An exceptional collection of bundles on G is given by

$$\tau = (\begin{smallmatrix} U \\ Q^* \end{smallmatrix}, \mathcal{O}_G, \mathcal{O}_G(1), \mathcal{O}_G(2), \mathcal{O}_G(3)).$$

PROOF. The exceptionality of the bundles in τ is proved in 2.1. The tables in 2.2 give all the conditions on the Ext's, except

$$^i\langle U \mid Q^* \rangle = {}^i\langle Q^* \mid U \rangle = 0, \quad \text{for } i \geqslant 0.$$

Observe that the first of these equalities is satisfied because

$$U^* \otimes Q^* = U(1) \otimes Q(-1) = U \otimes Q.$$

Consider the bundle

$$U \otimes U = U^* \otimes U^* \otimes \mathcal{O}_G(-2) = S^2 U^*(-2) \oplus \mathcal{O}_G(-1).$$

It follows from (2.1.3) that $H^i(U \otimes U) = 0$, for $i \geqslant 0$. Using this fact, we get what we need from (2.1.4) $\otimes U$.

2.4. Definition.

An infinite collection of bundles over G

$$\left(\ldots, E_{5k-1}, \frac{E_{5k}^1}{E_{5k}^2}, E_{5k+1}, E_{5k+2}, \ldots \right), \tag{2.4.1}$$

where $k \in \mathbf{Z}$, is *periodic* if $E_m = E_{m+5}(-4)$, for $m \neq 5k$, and $E_{5k}^j = e_{5(k+1)}^j(-4)$, for $j = 1, 2$.

A *foundation* of a periodic collection is any subcollection with five successive lower indices.

A periodic collection is called *exceptional* if all its foundations are exceptional.

2.5. Proposition.

The following periodic collection of bundles on G is exceptional:

$$\left(\ldots, \mathcal{O}_G(-5), \frac{U(-4)}{Q^*(-4)}, \mathcal{O}_G(-4), \mathcal{O}_G(-3), \mathcal{O}_G(-2), \mathcal{O}_G(-1), \ldots \right) \tag{2.5.1}$$

PROOF. This follows from Proposition 2.3 and the tables in 2.2.

From Remark 1.5, we can identify the periodic collection (2.4.1) with the collection

$$\sigma = (\ldots, E_m, \ldots), \quad m \in \mathbf{Z},$$

where $E_{5k} = E_{5k}^1 \oplus E_{5k}^2$ for all $k \in \mathbf{Z}$. The following conditions are sufficient to make σ a helix:

i) There exist four left shifts for five neighbouring bundles in σ.

ii) The resulting collections and σ itself, are exceptional (when written in the form (2.4.1)).

iii) $L_\sigma^{(4)} E_m = E_m(-4)$.

In the next section we shall verify these conditions for

$$\sigma = (\ldots, \mathcal{O}_G(-5), (U \oplus Q^*)(-4), \mathcal{O}_G(-4), \ldots).$$

3. Mutations of the Collection σ.

3.1. Preliminary Observations.

In this subsection, we obtain some results by considering fiberwise morphisms of vector bundles on G.

Let L be a point in $G(2, V)$ and $\{x, y\}$ a basis for L. Set $p = x \wedge y \in W$. Under the Plücker embedding, L maps to the line $\langle p \rangle \in \mathbf{P}(W)$.

Choose a basis $\zeta_1, \zeta_2, \zeta_3, \zeta_4$ for V and the dual basis $\zeta^1, \zeta^2, \zeta^3, \zeta^4$ for V^*. Fix a basis for W:

$$e_1 = \zeta_1 \wedge \zeta_2, \quad e_2 = \zeta_1 \wedge \zeta_3, \quad e_3 = \zeta_1 \wedge \zeta_4,$$
$$e_4 = \zeta_3 \wedge \zeta_4, \quad e_5 = \zeta_4 \wedge \zeta_2, \quad e_6 = \zeta_2 \wedge \zeta_3;$$

and let $e^1, \ldots, e^6$ be the dual basis in W^*. Let x and y have coordinates x^i and y^i with respect to the basis ζ_i and p have coordinates p^j in the basis e_j. Note that

$$p^j = \det \begin{pmatrix} x^k & y^k \\ x^l & y^l \end{pmatrix}, \quad \text{where } e_j = \zeta_k \wedge \zeta_l.$$

Note also that the coordinates p^j satisfy the relation

$$p^1 p^4 + p^2 p^5 + p^3 p^6 = 0, \tag{3.1.1}$$

which is just the equation of G in $\mathbf{P}^5$.

The fibre of the bundle $\mathcal{O}_G(d)$ over $\langle p \rangle \in G$ is given by

$$\mathcal{O}_G(d)_{\langle p \rangle} = \begin{cases} \underbrace{\langle p \rangle \otimes \cdots \otimes \langle p \rangle}_{-d \text{ times}} & \text{for } d < 0 \\ \mathbf{C} & \text{for } d = 0 \\ \underbrace{\langle p \rangle^* \otimes \cdots \otimes \langle p \rangle^*}_{d \text{ times}} & \text{for } d > 0. \end{cases}$$

Throughout this section, $\mathcal{O}_G(d)_{\langle p \rangle}$ is identified with $\mathbf{C}$, using the basis $p \otimes \cdots \otimes p$, for $d < 0$, and $p^* \otimes \cdots \otimes p^*$, for $d > 0$, where $p^*(p) = 1$.

If M is any complex vector space, then the fibre of the bundle $M \otimes \mathcal{O}_G(d)$ over a point $\langle p \rangle \in G$ will be identified with M using the isomorphism $\mathcal{O}_G(d)_{\langle p \rangle} \cong \mathbf{C}$ already established, i.e. $M \otimes \mathcal{O}_G(d)_{\langle p \rangle} \cong M \otimes \mathbf{C} = M$.

A section s of the bundle $\mathcal{O}_G(1)$ is determined by a covector $h = \sum_{j=1}^6 h_j e^j \in W^*$. In the fibre over $\langle p \rangle$, such a section is determined by $1 \mapsto h|_{\langle p \rangle} \in \langle p \rangle^*$. Since $e^j|_{\langle p \rangle} = p^j p^*$, we can interpret the section s in the fibre over $\langle p \rangle$ as a map $\mathbf{C} \to \mathbf{C} : 1 \mapsto \sum h_j p^j$

or simply as the linear form $\sum h_j p^j$. Note that any section s of a bundle $M \otimes \mathcal{O}_G(1)$ has a fiberwise interpretation as a map $\mathbf{C} \to M : 1 \mapsto \sum p^j t_j$, where $t_j \in M$ does not depend on p. Henceforth, we shall simply write $s = \sum p^j t_j$.

Let $v = \sum v^i \zeta_i \in L$. The fact that

$$\mathrm{rk} \begin{pmatrix} v^1 & x^1 & y^1 \\ v^2 & x^2 & y^2 \\ v^3 & x^3 & y^3 \\ v^4 & x^4 & y^4 \end{pmatrix} = 2.$$

yields the following relations between v^i and p^i:

$$\begin{aligned}
v^1 p^6 - v^2 p^2 + v^3 p^1 &= 0, \\
-v^1 p^5 - v^2 p^3 + v^4 p^1 &= 0, \\
v^1 p^4 - v^3 p^3 + v^4 p^2 &= 0, \\
v^2 p^6 + v^3 p^5 + v^4 p^6 &= 0,
\end{aligned} \tag{3.1.2}$$

In this section, we identify the space $(V/L)^*$ with its image under the natural inclusion $(V/L)^* \to V^*$. Let $f = \sum f_i \zeta^i \in (V/L)^*$. Then $f(x) = \sum x^i f_i = 0$ and $f(y) = \sum y^i f_i = 0$. A little calculation yields the four relations

$$\begin{aligned}
f_2 p^1 + f_3 p^2 + f_4 p^3 &= 0, \\
-f_1 p^1 + f_3 p^6 - f_4 p^5 &= 0, \\
f_1 p^2 + f_2 p^6 - f_4 p^4 &= 0, \\
f_1 p^3 - f_2 p^5 + f_3 p^4 &= 0.
\end{aligned} \tag{3.1.3}$$

A morphism of vector bundles $U \to \mathcal{O}_G$ is given by some covector $f = \sum f_i \zeta^i \in V^*$. In the fibre over a point L such a morphism is a restriction $f|_L : L \to \mathbf{C} : v \mapsto \sum f_i v^i$. A morphism of bundles $Q^* \to \mathcal{O}_G$ is determined by some vector $w = \sum w^i \zeta_i \in V$ by the rule $g \mapsto \sum g_i w^i$, where $g = \sum g_i \zeta^i \in (V/L)^* = Q_L^*$.

A morphism $U \to M \otimes \mathcal{O}_G$ has a fiberwise form very similar to that of a section of the bundle $M \otimes \mathcal{O}_G(1)$, i.e. $v \mapsto \sum v^i t_i$, for some $t_i \in M$. Similarly, a morphism $Q^* \to M \otimes \mathcal{O}_G$ has the form $g \mapsto \sum g_i t_i$, for some $t_i \in M$.

Choosing a basis for V establishes an isomorphism $\Lambda^4(V) \cong \mathbf{C} : \zeta_1 \wedge \zeta_2 \wedge \zeta_3 \wedge \zeta_4 \mapsto 1$ and an isomorphism $\Lambda^3(V) \cong V^*$, in which the three-vector $v_1 \wedge v_2 \wedge v_3$ is identified with the linear form $w \mapsto v_1 \wedge v_2 \wedge v_3 \wedge w \in \Lambda^4(V) \cong \mathbf{C}$. Note that any element of $(V/L)^*$ can be represented as $x \wedge y \wedge w$ for some $W \in V$.

Note finally that a morphism $\mathcal{O}_G(-1) \to U$ is given by some covector $f \in V^*$ and has the form of a contraction

$$p = x \wedge y \mapsto \langle f, x \wedge y \rangle = f(y)x - f(x)y,$$

while a morphism $\mathcal{O}_G(-1) \to Q^*$ is given by some vector $w \in V$ and has the form

$$p = x \wedge y \mapsto x \wedge y \wedge w \in (V/L)^*.$$

3.2. The Four Left Shifts of the Bundle $\mathcal{O}_G(3)$ in the Collection σ.

To shorten the notation, set $A_i = L_\sigma^{(i)} \mathcal{O}_G(3)$ for $i = 1, \dots, 4$.

Consider the exact sequence on $\mathbf{P}^5 = \mathbf{P}(W)$ dual to the Euler sequence

$$0 \to \Omega\mathbf{P}^5 \to W^* \otimes \mathcal{O}_{\mathbf{P}^5}(-1) \to \mathcal{O}_{\mathbf{P}^5} \to 0. \tag{3.2.1}$$

If we restrict this to G, we have

$$0 \to \Omega \to W^* \otimes \mathcal{O}_G(-1) \xrightarrow{\varphi_1} \mathcal{O}_G \to 0, \tag{3.2.2}$$

where Ω henceforth denotes the bundle $\Omega\mathbf{P}^5|_G$. We will also write $T = T\mathbf{P}^5|_G$.

The sequence (3.2.2) $\otimes \mathcal{O}_G(3)$ gives the first left shift of the bundle $\mathcal{O}_G(3)$ in σ, i.e. $A_1 = \Omega(3)$.

Given the identifications made in 3.1, we interpret the canonical morphism φ_1 and its twists $\varphi_1(d)$ in a similar way, i.e. over $\langle p \rangle \in G$, φ_1 induces the map on the fibres $W^* \to \mathbf{C}$ given by $p : e^j \mapsto p^j$.

Notice that $A_1 = \Omega(3)$ is an exceptional bundle on G. Indeed, from (3.2.2) it follows that $H^1(\Omega) = \mathbf{C}$ and $H^i(\Omega) = 0$ for $i \neq 1$; from (3.2.2) $\otimes \mathcal{O}_G(1)$ we get $H^i(\Omega(1)) = 0$ for $i \geqslant 0$; and from (3.2.2)$^* \otimes \Omega$ we get $H^0(T \otimes \Omega) = \mathbf{C}$ and $H^i(T \otimes \Omega) = 0$ for $i \geqslant 1$.

The canonical map

$$\mathcal{O}_G(1) \otimes {}^0\langle \mathcal{O}_G(1) \mid \Omega(3) \rangle \xrightarrow{\text{can}} \Omega(3)$$

is an epimorphism if and only if the canonical map $\mathcal{O}_G \otimes H^0(\Omega(2)) \xrightarrow{\text{can}} \Omega(2)$ is an epimorphism, i.e. $\Omega(2)$ is generated by its global sections. Let us interpret the space $H^0(\Omega(2))$. The sequence (3.2.1) $\otimes \mathcal{O}_{\mathbf{P}^5}(2)$ is

$$0 \to \Omega\mathbf{P}^5(2) \to W^* \otimes \mathcal{O}_{\mathbf{P}^5}(1) \to \mathcal{O}_{\mathbf{P}^5}(2) \to 0. \tag{3.2.3}$$

The long exact sequence, induced by (3.2.3) on cohomology, is

$$0 \to \Lambda^2 W^* \to W^* \otimes W^* \to S^2 W^* \to 0 \to 0 \to \cdots .$$

Let $\mathfrak{g} \in S^2 W^*$ be the equation for G in $\mathbf{P}^5$. Then $H^0(\mathcal{O}_G(2)) = S^2 W^*/\langle \mathfrak{g} \rangle$. Restricting (3.2.3) to G gives

$$0 \to \Omega(2) \to W^* \otimes \mathcal{O}_G(1) \to \mathcal{O}_G(2) \to 0. \tag{3.2.4}$$

The long exact cohomology sequence for (3.2.4) has the form

$$0 \to \Lambda^2 W^* \oplus \langle \mathfrak{g} \rangle \to W^* \otimes W^* \to S^2 W^*/\langle \mathfrak{g} \rangle \to 0 \to 0 \to \cdots ,$$

i.e. $H^0(\Omega(2)) = \Lambda^2 W^* \oplus \langle \mathfrak{g} \rangle$ is a subspace of $W^* \otimes W^*$ and $h^0(\Omega(2)) = 16$.

The bundle $\Omega\mathbf{P}^5(2)$ is generated by its global sections. This follows from the exact sequence

$$0 \to \Lambda^2 \Omega\mathbf{P}^5(2) \to \Lambda^2 W^* \otimes \mathcal{O}_{\mathbf{P}^5} \to \Omega\mathbf{P}^5(2) \to 0, \tag{3.2.5}$$

which is obtained by decomposing the twisted Koszul complex

$$\begin{aligned}
0 \to \mathcal{O}_{\mathbf{P}^5}(-4) &\to \Lambda^5 W^* \otimes \mathcal{O}_{\mathbf{P}^5}(-3) \to \Lambda^4 W^* \otimes \mathcal{O}_{\mathbf{P}^5}(-2) \to \\
&\to \Lambda^3 W^* \otimes \mathcal{O}_{\mathbf{P}^5}(-1) \to \Lambda^2 W^* \otimes \mathcal{O}_{\mathbf{P}^5} \to W^* \otimes \mathcal{O}_{\mathbf{P}^5}(1) \to \mathcal{O}_{\mathbf{P}^5}(2) \to 0
\end{aligned} \tag{3.2.6}$$

to short exact sequences. Hence, on G the maps $\Lambda^2 W^* \otimes \mathcal{O}_{\mathbf{P}^5} \to \Omega(2)$ and

$$\left(\Lambda^2 W^* \oplus \langle \mathfrak{g} \rangle \right) \otimes \mathcal{O}_{\mathbf{P}^5} \to \Omega(2)$$

are epimorphic.

Thus we can determine the second left shift of $\mathcal{O}_G(3)$ from

$$0 \to A_2 \to H^0(\Omega(2)) \otimes \mathcal{O}_G(1) \xrightarrow{\varphi_2} \Omega(3) \to 0. \tag{3.2.7}$$

Let us interpret the canonical morphism φ_2. Choose a basis in $H^0(\Omega(2)) = \Lambda^2 W^* \oplus \langle \mathfrak{g} \rangle$, composed of $e^{ij} = e^i \wedge e^j$, for $1 \leqslant i < j \leqslant 6$, and

$$\mathfrak{g} = \frac{1}{2}\left(e^1 \otimes e^4 + e^2 \otimes e^5 + e^3 \otimes e^6 + e^4 \otimes e^1 + e^5 \otimes e^2 + e^6 \otimes e^3 \right).$$

We can consider φ_2 to be a map into $W^* \otimes \mathcal{O}_G(2)$, whose image is $\Omega(3)$, which is a subbundle of $W^* \otimes \mathcal{O}_G(2)$. Taking into account the identifications made in 3.1, we have, at the fibre over $\langle p \rangle$, the maps

$$e^{ij} \mapsto p^j e^i - p^i e^j,$$

$$\mathfrak{g} \mapsto \frac{1}{2}\left(p^1 e^4 + p^2 e^5 + p^3 e^6 + p^4 e^1 + p^5 e^2 + p^6 e^3 \right).$$

The images of the basis vectors of $H^0(\Omega(2))$ are sections of $\Omega(2)$, or, in the notation of 3.1, sections of $W^* \otimes \mathcal{O}_G(1)$ which map to zero under $\varphi_1(2)$.

3.2.3 PROPOSITION. *The bundle A_2 is exceptional.*

PROOF. From $(3.2.7) \otimes \mathcal{O}_G(-1)$ it follows that $H^i(A_2(-1)) = 0$ for $i \geqslant 0$; from $(3.2.2)^*$ that $H^i(T(-2)) = 0$ for $i \geqslant 0$. Since Ω is exceptional, it follows from $(3.2.7) \otimes T(-3)$ that $H^1(A_2 \otimes T(-3)) \cong \mathbb{C}$ and $H^i(A_2 \otimes T(-3)) = 0$ for $i \neq 1$. Finally, from $(3.2.7)^* \otimes A_2$ it follows that ${}^0\langle A_2 \mid A_2 \rangle \cong \mathbb{C}$ and ${}^i\langle A_2 \mid A_2 \rangle = 0$ for $i \geqslant 1$.

3.2.4 LEMMA. *If the vector bundle E over G satisfies $H^i(\Omega(3) \otimes E) = 0$, $H^i(\Lambda^2\Omega(3) \otimes E) = 0$ and $H^i(E(1)) = 0$ for $i \geqslant 1$, then $H^i(E \otimes A_2) = 0$ for $i \geqslant 1$.*

PROOF. Consider the sequences $(3.2.5) \otimes E(1)$

$$0 \to \Lambda^2\Omega(3) \otimes E \to \Lambda^2 W^* \otimes E(1) \to \Omega(3) \otimes E \to 0$$

and $(3.2.7) \otimes E$

$$0 \to A_2 \otimes E \to H^0(\Omega(2)) \otimes E(1) \to \Omega(3) \otimes E \to 0.$$

The long exact cohomology sequence for $(3.2.7) \otimes E$ takes the form

$$0 \to H^0(A_2 \otimes E) \to (\Lambda^2 W^* \oplus \langle \mathfrak{g} \rangle) \otimes H^0(E(1)) \xrightarrow{\gamma} H^0(\Omega(3) \otimes E) \to$$
$$\to H^1(A_2 \otimes E) \to 0 \to 0 \to \cdots.$$

From the conditions of the lemma it follows that, in $(3.2.5) \otimes E(1)$, the induced map on sections $\Lambda^2 W^* \otimes H^0(E(1)) \to H^0(\Omega(3) \otimes E)$ is an epimorphism, which means that the map γ is also an epimorphism. Consequently, $H^1(A_2 \otimes E) = 0$, completing the proof.

The following short exact sequences, obtained by partitioning the long exact sequence $(3.2.6)|_G \otimes \mathcal{O}_G(1)$, are often helpful in verifying the conditions of the lemma:

$$0 \to \mathcal{O}_G(-3) \to \Lambda^5 W^* \otimes \mathcal{O}_G(-2) \to \Lambda^4\Omega(3) \to 0 \qquad (3.2.8)$$

$$0 \to \Lambda^4\Omega(3) \to \Lambda^4 W^* \otimes \mathcal{O}_G(-1) \to \Lambda^3\Omega(3) \to 0 \qquad (3.2.9)$$

$$0 \to \Lambda^3\Omega(3) \to \Lambda^3 W^* \otimes \mathcal{O}_G \to \Lambda^2\Omega(3) \to 0 \qquad (3.2.10)$$

$$0 \to \Lambda^2\Omega(3) \to \Lambda^2 W^* \otimes \mathcal{O}_G(1) \to \Omega(3) \to 0 \qquad (3.2.11)$$

3.2.5 PROPOSITION. $h^0(A_2) = 26$ *and* $H^i(A_2) = 0$ *for* $i \geqslant 1$.

PROOF. From $(3.2.8)$–$(3.2.11)$ it follows that $h^0(\Omega(3)) = 70$ and that the conditions of Lemma 3.2.4 are satisfied for $\mathcal{O}_G$. Hence, $H^i(A_2) = 0$ for $i \geqslant 1$ and, from $(3.2.7)$, we get $h^0(A_2) = 26$. $\square$

Consider all possible three-element subsets $\{i, j, k\}$ of $\{1, \ldots, 6\}$, where $1 \leqslant i < j < k \leqslant 6$. Number these subsets lexicographically and let N_{ijk} be the number of the subset $\{i, j, k\}$, e.g. $N_{123} = 1$ and $N_{456} = 20$. Consider the following 26 sections of the bundle $H^0(\Omega(2)) \otimes \mathcal{O}_G(1)$:

$$S_{N_{ijk}} = p^k e^{ij} - p^j e^{ik} + p^i e^{jk}, \quad (1 \leqslant i < j < k \leqslant 6),$$

$$S_{21} = p^1 \mathfrak{g} + \frac{1}{2} p^1 (e^{14} + e^{25} + e^{36}) + p^5 e^{12} + p^6 e^{13},$$

$$S_{22} = p^2 \mathfrak{g} + \frac{1}{2} p^2 (e^{14} + e^{25} + e^{36}) + p^6 e^{23} - p^4 e^{12},$$

$$S_{23} = p^3 \mathfrak{g} + \frac{1}{2} p^3 (e^{14} + e^{25} + e^{36}) - p^4 e^{13} - p^5 e^{23},$$

$$S_{24} = p^4 \mathfrak{g} - \frac{1}{2} p^4 (e^{14} + e^{25} + e^{36}) + p^2 e^{45} + p^3 e^{46},$$

$$S_{25} = p^5 \mathfrak{g} - \frac{1}{2} p^5 (e^{14} + e^{25} + e^{36}) - p^1 e^{45} + p^3 c^{56},$$

$$S_{26} = p^6 \mathfrak{g} - \frac{1}{2} p^6 (e^{14} + e^{25} + e^{36}) - p^1 e^{46} - p^2 e^{56}.$$

One can easily see that $S_1, \ldots, S_{26}$ are sections of A_2, i.e. they are all mapped to zero by φ_2. Furthermore, these sections are linearly independent. Indeed, suppose that there are some $\mu_i \in \mathbb{C}$ such that $\mu_1 S_1 + \cdots + \mu_{26} S_{26} = 0$ at every point $\langle p \rangle \in G$. Substituting for p each of e_1, e_2, e_3 and e_4 in turn, we deduce that $\mu_1 = \cdots = \mu_{26} = 0$. Therefore $S_1, \ldots, S_{26}$ is a basis for $H^0(A_2)$.

We now prove that the canonical map $H^0(A_2) \otimes \mathcal{O}_G \xrightarrow{\text{can}} A_2$ is an epimorphism. Using the homogeneity of $G(2, V)$ under the action of the group $GL(V)$ and the exceptionality of the bundle A_2, we see that it is sufficient to check that, over a fixed point $\langle p \rangle \in G$, the rank of the system of vectors $\{S_i|_{\langle p \rangle} : i = 1, \ldots, 26\}$, in the fibre of the bundle $H^0(\Omega(2)) \otimes \mathcal{O}_G(1)$ at $\langle p \rangle$, is equal to $\text{rk}(A_2) = 11$. This is easily done for $p = e_1$.

Therefore the third left shift of the bundle $\mathcal{O}_G(3)$ is determined by

$$0 \to A_3 \to H^0(A_2) \otimes \mathcal{O}_G \xrightarrow{\text{can}} A_2 \to 0. \qquad (3.2.12)$$

3.2.6 PROPOSITION. *The bundle A_3 is exceptional.*

PROOF. From $(3.2.2)^* \otimes \mathcal{O}_G(-3)$ it follows that $H^i(T(-3)) = 0$ for $i \geqslant 0$; from $(3.2.7)^*$ that $H^i(A^*) = 0$ for $i \geqslant 0$; and from $(3.2.12) \otimes A_2^*$ that $H^1(A_2^* \otimes A_3) \cong \mathbb{C}$ and $H^i(A_2 \otimes A_3) = 0$ for $i \neq 1$. Finally, from $(3.2.12)^* \otimes A_3$, we get ${}^0\langle A_3 \mid A_3 \rangle \cong \mathbb{C}$ and ${}^i\langle A_3 \mid A_3 \rangle = 0$ for $i \geqslant 1$.

3.2.7 LEMMA. *If a bundle E over G is generated by its global sections, i.e. there exists a short exact sequence*

$$0 \to B \to H^0(E) \otimes \mathcal{O}_G \to E \to 0, \qquad (3.2.13)$$

and if also $H^i(E) = 0$ for $i \geqslant 1$ and $H^i(B \otimes A_2) = 0$ for $i \geqslant 1$, then $H^i(E \otimes A_3) = 0$ for $i \geqslant 1$.

PROOF. From $(3.2.13) \otimes A_2$ we see that $H^i(E \otimes A_2) = 0$ for $i \geqslant 1$. In particular, the induced map on sections $H^0(E) \otimes H^0(A_2) \to H^0(E \otimes A_2)$ is an epimorphism. The same map is induced by the bundle morphism $H^0(A_2) \otimes E \to A_2 \otimes E$ in the short exact sequence $(3.2.12) \otimes E$, from which we find that $H^i(A_3 \otimes E) = 0$ for $i \geqslant 1$, as required.

PROPOSITION 3.2.8. $\dim \operatorname{Hom}(U, A_3) = \dim \operatorname{Hom}(Q^*, A_3) = 4$.

PROOF. Note that $\dim \operatorname{Hom}(U, A_3) = h^0(U^* \otimes A_3)$. From $(3.2.8) \otimes Q^*$–$(3.2.11) \otimes Q^*$ it follows that $h^0(Q^* \otimes \Omega(3)) = 60$ and that Q^* satisfies the conditions of Lemma 3.2.4. Hence, $H^i(A_2 \otimes Q^*) = 0$ for $i \geqslant 1$. From $(3.2.7) \otimes Q^*$ we get $h^0(A_2 \otimes Q^*) = 4$. Applying Lemma 3.2.7 to the bundle U^* (c.f. (2.1.5)), we have $H^i(U^* \otimes A_3) = 0$ for $i \geqslant 1$. From $(2.1.5) \otimes A_2$ we find that $h^0(U^* \otimes A_2) = 100$; and from $(3.2.12) \otimes U^*$ that $h^0(U^* \otimes A_3) = 4$.

Since $h^i(U^*(d)) = h^i(Q(d))$, we can repeat the above argument changing Q^* to U, U^* to Q and (2.1.5) to (2.1.4). Thus, we get $\dim \operatorname{Hom}(Q^*, A_3) = 4$, completing the proof.

Choose an arbitrary vector $v = \sum v^i \zeta_i \in L = U_L$ and an arbitrary covector $f = \sum f_i \zeta^i \in (V/L)^* = Q_L^*$. Consider the bundle morphisms $\alpha_1, \alpha_2, \alpha_3, \alpha_4 : U \to H^0(A_2) \otimes \mathcal{O}_G$ defined in the fibre over L as follows:

$$\alpha_1 : v \mapsto -v^1 S_{20} + v^2 S_{24} + v^3 S_{25} + v^4 S_{26},$$

$$\alpha_2 : v \mapsto v^1(S_{24} - S_{14} - S_{18}) + v^2 S_{11} - v^3(S_5 + S_{23}) + v^4(S_{22} + S_2),$$

$$\alpha_3 : v \mapsto v^1(S_{25} + S_3 - S_{19}) + v^2(S_{12} + S_{23}) - v^3 S_6 + v^4(S_3 - S_{21}),$$

$$\alpha_4 : v \mapsto v^1(S_{26} + S_9 + S_{16}) + v^2(S_{13} - S_{22}) - v^3(S_{21} - S_7) + v^4 S_4.$$

Consider also the bundle morphisms $\beta_1, \beta_2, \beta_3, \beta_4 : Q^* \to H^0(A_2) \otimes \mathcal{O}_G$ defined on the fibre over L as follows:

$$\beta_1 : f \mapsto f_1 S_1 - f_2 S_{21} - f_3 S_{22} - f_4 S_{23},$$

$$\beta_2 : f \mapsto f_1(S_{21} - S_3 - S_7) + f_2 S_{10} - f_3(S_{26} - S_9) + f_4(S_{25} + S_8),$$

$$\beta_3 : f \mapsto f_1(S_{22} + S_2 - S_{13}) + f_2(S_{26} + S_{16}) - f_3 S_{15} + f_4(S_{14} - S_{24}),$$

$$\beta_4 : f \mapsto f_1(S_{23} + S_{12} + S_5) + f_2(S_{19} - S_{25}) + f_3(S_{24} - S_{18}) + f_4 S_{17}.$$

Using identities (3.1.2) and (3.1.3), one can easily check that the images of any $v \in U_L$ and $f \in Q_L^*$ lie in the kernel of the canonical map $H^0(A_2) \otimes \mathcal{O}_G \xrightarrow{\text{can}} A_2$, i.e. the maps α_i and β_i take values in the bundles A_3. Also, it is easy to see that the α_i and the β_i are both linearly independent. With Proposition 3.2.8, this means that the α_i form a basis of $\operatorname{Hom}(U, A_3)$ and the β_i form a basis of $\operatorname{Hom}(Q^*, A_3)$.

Consequently, $\alpha_i \otimes x$, $\alpha_k \otimes y$ is a basis in the fibre over L of the bundle $\mathrm{Hom}(U, A_3) \otimes U$. If we choose a basis g, h for Q_L^*, then we get a basis $\beta_j \otimes g$, $\beta_l \otimes h$ in the fibre over L of the bundle $\mathrm{Hom}(Q^*, A_3) \otimes Q^*$.

Since $G(2, V)$ is homogeneous with respect to the action of the group $GL(V)$ and since the bundles A_3, U and Q^* are exceptional, it is sufficient that the map

$$\begin{array}{c} U \otimes {}^0\langle U \mid A_3 \rangle \\ \oplus \\ Q^* \otimes {}^0\langle Q^* \mid A_3 \rangle \end{array} \longrightarrow A_3$$

is an epimorphism at one point. It is easy to see that the rank of the system of vectors $\alpha_{i L_0}(\zeta_1)$, $\alpha_{k L_0}(\zeta_2)$, $\beta_{j L_0}(\zeta^3)$, $\beta_{l L_0}(\zeta^4)$, in the fibre over $L_0 = \langle \zeta_1, \zeta_2 \rangle$ of the bundle $H^0(a_2) \otimes \mathcal{O}_G$ is equal to fifteen, i.e. $\mathrm{rk}(A_3)$. Hence, on G there exists an exact sequence of vector bundles

$$0 \to A_4 \to \begin{array}{c} U \otimes {}^0\langle U \mid A_3 \rangle \\ \oplus \\ Q^* \otimes {}^0\langle Q^* \mid A_3 \rangle \end{array} \to A_3 \to 0. \tag{3.2.14}$$

From this we immediately have $\mathrm{rk}\, A_4 = 1$. Calculations from the short exact sequences $(3.2.2) \otimes \mathcal{O}_G(3)$, $(3.2.7)$, $(3.2.12)$, $(3.2.14)$ show that $c_1(A_4) = -1$. Hence, $A_4 \cong \mathcal{O}_G(-1)$, i.e. $L_\sigma^{(4)} \mathcal{O}_G(3) = \mathcal{O}_G(-1)$. $(3.2.14)$ can be spliced together to form the long sequence

$$0 \to \mathcal{O}_G(-1) \to \begin{array}{c} U \otimes {}^0\langle U \mid A_3 \rangle \\ \oplus \\ Q^* \otimes {}^0\langle Q^* \mid A_3 \rangle \end{array} \to H^0(A_2) \otimes \mathcal{O}_G \to \tag{3.2.15}$$

$$\to H^0(\Omega(2)) \otimes \mathcal{O}_G(1) \to W^* \otimes \mathcal{O}_G(2) \to \mathcal{O}_G(3) \to 0$$

If one were able to obtain this sequence without having to make such direct calculation, this might make its meaning more clear.

3.3. The Four Left Shifts of the Bundle $\mathcal{O}_G(2)$ in the Collection σ.

The first two shifts are already known from the previous sections:

$$L_\sigma^{(1)} \mathcal{O}_G(2) = \Omega(2) \quad \text{and} \quad L_\sigma^{(2)} \mathcal{O}_G(2) = A_2(-1).$$

They appear in the short exact sequences $(3.2.4)$ and $(3.2.7) \otimes \mathcal{O}_G(-1)$.

Consider the canonical map

$$\begin{array}{c} U \otimes {}^0\langle U \mid A_2(-1) \rangle \\ \oplus \\ Q^* \otimes {}^0\langle Q^* \mid A_2(-1) \rangle \end{array} \longrightarrow A_2(-1), \tag{3.3.1}$$

where $\dim{}^0\langle U \mid A_2(-1)\rangle = \dim{}^0\langle Q^* \mid A_2(-1)\rangle = h^0(Q^* \otimes A_2) = 4$, as is calculated in the proof of Proposition 3.2.8.

As in the previous section, let $v = \sum v^i\zeta_i \in L = U_L$ and $f = \sum f_i\zeta^i \in (V/L)^* = Q_L^*$. Consider the bundle morphisms $\lambda_1, \ldots, \lambda_4 : U \to H^0\big(\Omega(2)\big) \otimes \mathcal{O}_G$, which are defined in the fibre over L as follows:

$$\lambda_1 : v \mapsto v^1\big(\mathfrak{g} + \frac{1}{2}(e^{14} + e^{25} + e^{36})\big) + v^2 e^{23} - v^3 e^{13} - v^4 e^{12}$$

$$\lambda_2 : v \mapsto v^1 e^{56} + v^2\big(\mathfrak{g} + \frac{1}{2}(e^{14} - e^{25} - e^{36})\big) - v^3 e^{15} - v^4 e^{16}$$

$$\lambda_3 : v \mapsto v^1 e^{46} + v^2 e^{24} + v^3\big(\mathfrak{g} + \frac{1}{2}(e^{25} - e^{36} - e^{14})\big) + v^4 e^{26}$$

$$\lambda_4 : v \mapsto v^1 e^{45} - v^2 e^{34} - v^3 e^{35} + v^4\big(\mathfrak{g} + \frac{1}{2}(e^{36} - e^{14} - e^{25})\big).$$

Consider also the bundle morphisms $\eta_1, \ldots, \eta_4 : Q^* \to H^0\big(\Omega(2)\big) \otimes \mathcal{O}_G$, defined in the fibre over L as follows:

$$\eta_1 : f \mapsto f_1\big(\mathfrak{g} - \frac{1}{2}(e^{14} + e^{25} + e^{36})\big) + f_2 e^{56} - f_3 e^{46} + f_4 e^{45}$$

$$\eta_2 : f \mapsto -f_1 e^{23} + f_2\big(\mathfrak{g} - \frac{1}{2}(e^{14} - e^{25} - e^{36})\big) - f_3 e^{24} - f_4 e^{34}$$

$$\eta_3 : f \mapsto -f_1 e^{13} + f_2 e^{15} - f_3\big(\mathfrak{g} - \frac{1}{2}(e^{25} - e^{36} - e^{14})\big) + f_4 e^{35}$$

$$\eta_4 : f \mapsto f_1 e^{12} + f_2 e^{16} + f_3 e^{26} - f_4\big(\mathfrak{g} - \frac{1}{2}(e^{36} - e^{14} - e^{25})\big).$$

Using identities (3.1.2) and (3.1.3), it is easy to check that the images of any $v \in U_L$ and $f \in Q_L^*$ lie in the kernel of the canonical map $H^0\big(\Omega(2)\big) \xrightarrow{\text{can}} \Omega(2)$, i.e. the maps λ_i and η_i take values in the bundle $A_2(-1)$. It is also easy to check that the λ_i are linearly independent and that the η_i are linearly independent. Consequently, the λ_i are a basis in the fibre over L of the bundle ${}^0\langle U \mid A_2(-1)\rangle$ and the η_i are a basis in the fibre over L of the bundle ${}^0\langle Q^* \mid A_2(-1)\rangle$. This means that $\{\lambda_i \otimes x, \lambda_k \otimes y\}$ is a basis in the fibre over L of the bundle $U \otimes {}^0\langle U \mid A_2(-1)\rangle$. As before, if we choose a basis g, h for Q_L^*, then we get a basis $\{\beta_j \otimes g, \beta_l \otimes h\}$ in the fibre over L of the bundle $Q^* \otimes {}^0\langle Q^* \mid A_2(-1)\rangle$.

Since $G(2, V)$ is homogeneous under the action of the group $GL(V)$ and since the bundles $A_2(-1)$, U and Q^* are exceptional, it is enough to check that the map (3.3.1) is an epimorphism at one point in G. Over $L_0 = \langle \zeta_1, \zeta_2\rangle$ the rank of the system of vectors $\lambda_{iL_0}(x)$, $\lambda_{kL_0}(y)$, $\eta_{jL_0}(g)$, $\eta_{lL_0}(h)$ is equal to $11 = \operatorname{rk} A_2(-1)$. Consequently, on G there is an exact sequence of bundles

$$0 \to K \to \begin{array}{c} U \otimes {}^0\langle U \mid A_2(-1)\rangle \\ \oplus \\ Q^* \otimes {}^0\langle Q^* \mid A_2(-1)\rangle \end{array} \to A_2(-1) \to 0, \qquad (3.3.2)$$

where $K = L_\sigma^{(3)}\mathcal{O}_G(2)$ and $\operatorname{rk} K = 5$.

3.3.1 PROPOSITION. $K \cong T(-2)$.

PROOF. We show that K can be inserted into the exact sequence

$$0 \to \mathcal{O}_G(-2) \xrightarrow{\gamma} W \otimes \mathcal{O}_G(-1) \to K \to 0,$$

where γ is the canonical inclusion, which has the fibrewise form $\mathbf{C} \to W : 1 \mapsto \sum p^j e_j = p$. Consider the morphism

$$\psi : W \otimes \mathcal{O}_G(-1) \longrightarrow \begin{array}{c} U \otimes {}^0\langle U \mid A_2(-1)\rangle \\ \oplus \\ Q^* \otimes {}^0\langle Q^* \mid A_2(-1)\rangle \end{array},$$

defined over a point L as follows:

$$e_1 \mapsto \left(\lambda_4 \otimes (y^3 x - x^3 y) + \lambda_3 \otimes (y^4 x - x^4 y), \eta_1 \otimes (x \wedge y \wedge \zeta_2) - \eta_2 \otimes (x \wedge y \wedge \zeta_1)\right)$$
$$e_2 \mapsto \left(\lambda_4 \otimes (x^2 y - y^2 x) + \lambda_2 \otimes (y^4 x - x^4 y), \eta_1 \otimes (x \wedge y \wedge \zeta_3) + \eta_3 \otimes (x \wedge y \wedge \zeta_1)\right)$$
$$e_3 \mapsto \left(\lambda_3 \otimes (x^2 y - y^2 x) + \lambda_2 \otimes (x^3 y - y^3 x), \eta_1 \otimes (x \wedge y \wedge \zeta_4) + \eta_4 \otimes (x \wedge y \wedge \zeta_1)\right)$$
$$e_4 \mapsto \left(\lambda_2 \otimes (y^1 x - x^1 y) + \lambda_1 \otimes (y^2 x - x^2 y), \eta_4 \otimes (x \wedge y \wedge \zeta_3) - \eta_3 \otimes (x \wedge y \wedge \zeta_4)\right)$$
$$e_5 \mapsto \left(\lambda_3 \otimes (x^1 y - y^1 x) + \lambda_1 \otimes (y^3 x - x^3 y), -\eta_2 \otimes (x \wedge y \wedge \zeta_4) - \eta_4 \otimes (x \wedge y \wedge \zeta_2)\right)$$
$$e_6 \mapsto \left(\lambda_4 \otimes (y^1 x - x^1 y) + \lambda_1 \otimes (y^4 x - x^4 y), \eta_2 \otimes (x \wedge y \wedge \zeta_3) + \eta_3 \otimes (x \wedge y \wedge \zeta_2)\right)$$

One can easily check that, over any point L, the images of the basis vectors $\psi_L(e_1)$, $\ldots, \psi_L(e_6)$ are mapped to zero by the canonical map (3.3.1), i.e. ψ maps $W \otimes \mathcal{O}_G(-1)$ to K. We now show that ψ is maps $W \otimes \mathcal{O}_G(-1)$ onto K, by checking that the fibrewise map

$$\psi_L : W \longrightarrow \begin{array}{c} U_L \otimes {}^0\langle U \mid A_2(-1)\rangle \\ \oplus \\ Q_L^* \otimes {}^0\langle Q^* \mid A_2(-1)\rangle \end{array}$$

has rank five on the fibre over L, for every L. Consider the map

$$\psi_L^+ : W \to U_L \otimes {}^0\langle U \mid A_2(-1)\rangle,$$

induced by ψ_L. We write down the matrix of the map ψ_L^+ with respect to the basis of $U_L \otimes {}^0\langle U \mid A_2(-1)\rangle$ given above:

	$\psi_L^+(e_1)$	$\psi_L^+(e_2)$	$\psi_L^+(e_3)$	$\psi_L^+(e_4)$	$\psi_L^+(e_5)$	$\psi_L^+(e_6)$
$\lambda_1 \otimes x$	0	0	0	y^2	y^3	y^4
$\lambda_1 \otimes y$	0	0	0	$-x^2$	$-x^3$	$-x^4$
$\lambda_2 \otimes x$	0	y^4	$-y^3$	y^1	0	0
$\lambda_2 \otimes y$	0	$-x^4$	x^3	$-x^1$	0	0
$\lambda_3 \otimes x$	y^3	0	$-y^2$	0	$-y^1$	0
$\lambda_3 \otimes y$	$-x^3$	0	x^2	0	x^1	0
$\lambda_4 \otimes x$	y^4	$-y^2$	0	0	0	y^1
$\lambda_4 \otimes y$	$-x^4$	x^2	0	0	0	$-x^1$

$$(3.3.3)$$

Since in all cases $\det \begin{pmatrix} x^i & y^i \\ x^j & y^j \end{pmatrix} \neq 0$ for $1 \leqslant i < j \leqslant 6$, we see that the matrix (3.3.3) always has a 5×5 submatrix with nonzero determinant. This means that $\operatorname{rk} \psi_L^+ \geqslant 5$, and hence $\operatorname{rk} \psi_L = 5$.

Note that, from identities (3.1.2) and (3.1.3), we have $\psi \circ \gamma = 0$. Therefore there is an exact sequence of bundles on G.

$$0 \to \mathcal{O}_G(-2) \xrightarrow{\gamma} W \otimes \mathcal{O}_G(-1) \xrightarrow{\psi} K \to 0.$$

This completes the proof of Proposition 3.3.1. It follows from this proof that ψ is the canonical map

$$\mathcal{O}_G(-1) \otimes {}^0 \langle \mathcal{O}_G(-1) \mid T(-2) \rangle \xrightarrow{\text{can}} T(-2),$$

i.e. $L_\sigma^{(4)} \mathcal{O}_G(2) = \mathcal{O}_G(-2)$.

As in the previous section, the short exact sequences $(3.2.4)^*$, $(3.3.2)$, $(3.2.7) \otimes \mathcal{O}_G(-1)$ and $(3.2.4)$ can be spliced together to form the long sequence

$$0 \to \mathcal{O}_G(-2) \to W \otimes \mathcal{O}_G(-1) \to \begin{matrix} U \otimes {}^0 \langle U \mid A_2(-1) \rangle \\ \oplus \\ Q^* \otimes {}^0 \langle Q^* \mid A_2(-1) \rangle \end{matrix} \to \qquad (3.3.4)$$

$$\to H^0(\Omega(2)) \otimes \mathcal{O}_G \to W^* \otimes \mathcal{O}_G(1) \to \mathcal{O}_G(2) \to 0.$$

3.4. The Left Shifts of the Bundles $\mathcal{O}_G(1)$, $\mathcal{O}_G$ and $U \oplus Q^*$.

Since $\sigma = \sigma^* \otimes \mathcal{O}_G(-1)$, decomposing the sequence $(3.3.4) \otimes \mathcal{O}_G(-1)$ (respectively $(3.2.15)^* \otimes \mathcal{O}_G(-1)$) into short exact sequences gives the four new left shifts of $\mathcal{O}_G(1)$ (resp. $\mathcal{O}_G$) in σ and proves the equality $L_\sigma^{(4)} \mathcal{O}_G(1) = \mathcal{O}_G(-3)$ (resp. $L_\sigma^{(4)} \mathcal{O}_G(1) = \mathcal{O}_G(-4)$).

From the short exact sequences $(2.1.4) \otimes \mathcal{O}_G(-1)$ and $(2.1.5) \otimes \mathcal{O}_G(-1)$, it follows that

$$L_\sigma^{(1)} U \oplus Q^* = U(-1) \oplus Q^*(-1) = (U \oplus Q^*)(-1).$$

The collection we get is equal to $\sigma \otimes \mathcal{O}_G(-1)$. Therefore, the four left shifts of $U \oplus Q^*$ in σ are defined and $L^{(4)} U \oplus W^* = (U \oplus Q^*)(-4)$.

It is straightforward to check that the collections obtained by making all possible shifts in σ are exceptional using the exact sequences in this paper and Lemmas 3.2.4 and 3.2.7. Thus we have the following result:

THEOREM. *The following exceptional collection of bundles on G is a helix.*

$$(\ldots, (U \oplus Q^*)(-4), \mathcal{O}_G(-4), \mathcal{O}_G(-3), \mathcal{O}_G(-2), \mathcal{O}_G(-1), U \oplus Q^*, \ldots)$$

References

[1] GORODENTSEV, A.L. & RUDAKOV, A.N., Exceptional Vector Bundles on Projective Space, *Duke Math. J.*, **54** (1987) 115–130.

[2] KAPRANOV, M.M., The Derived Category of Coherent Sheaves on a Quadric, *Funk. An.*, **20** (1986) 67

[3] RUDAKOV, A.N., Exceptional Bundles on a Quadric, *Math. USSR Isv.*, **33** (1989) 115–138.

Index